Anatomy & Development of the
GRAND PRIX
MOTORCYC

Tony Sakkis

Motorbooks International

First published in 1995 by Motorbooks International Publishers & Wholesalers, PO Box 2, 729 Prospect Avenue, Osceola, WI 54020 USA

Motorbooks International books are also available at discounts in bulk quantity for industrial or sales-promotional use. For details write to Special Sales Manager at the Publisher's address

Library of Congress Cataloging-in-Publication Data

Sakkis, Tony
Anatomy & development of the GP motorcycle / Tony Sakkis.
    p.   cm.
  Includes index.
  ISBN 0-87938-978-8 (pbk.)
  1. Motorcycles, Racing—History.   I. Title. II. Title: Anatomy and development of the Grand Prix motorcycle.
TL442.S25 1995
629.227'5—dc20   95-6116

On the front cover: Team Roberts Marlboro rider Daryll Beattie aboard the Yamaha YZR500, 1994. *Brian Nelson*

On the back cover: Left: 1994 World Champion Mick Doohan at speed. Right: ROC-framed Yamaha YZR500. Middle: John Surtees' World Championship-winning MV Agusta, 1956. Mick Walker collection Bottom: Six-piston calipers haul this Suzuki RGV500 to a stop.

Printed and bound in the United States of America

CONTENTS

ACKNOWLEDGMENTS

*I*n no other project have my feelings for my subject grown as much as with this one. It wasn't as if my enthusiasm for the sport grew any fonder as I learned about the subtleties of the Grand Prix motorcycling, but my respect for the people involved has shot through the roof.

Talking with World Champions—especially ones with talent such as these men have—is a humbling experience. But it has become so much more powerful to me since I realize the amount of humility *they* have. Almost without exception, every one of these guys is a gentleman.

Thanks to Mick Walker whose historical photography was invaluable.

Thanks to Mick Doohan, Wayne Gardner, Erv Kanemoto, Eddie Lawson, Niall MacKenzie, Randy Mamola, Kenny Roberts, Serge Rosset, and Kevin Schwantz. If I still have heroes after my years in the business, these are the guys.

Thanks also to Giacomo Agostini, Bud Aksland, Kevin Cameron, Martin Cavanaugh, Jeremy Ferguson, Dan Gurney, Nick Harris, Claudine Luyten, Robert MacLean, Leo Mehl, Tad Pilati, Bob Schultz, Danny Walker, Mike Webb, and Steve Whiteleock.

Special thanks to Suzuki PR man, Ty Van Hoydonk, who jumped through hoops to get me contacts, interviews, and general background information to make this book work. I couldn't have done it without him.

Finally, extra thanks to Kenny Roberts, Tad Pilati, and Team Marlboro Roberts who graciously allowed me to make a nuisance of myself two years straight. There is no finer organization in GP racing.

INTRODUCTION

At first, it's nothing more than a buzz. A metallic groan accompanied by an aerodynamic whistle which, as it gets closer, sounds more like a shriek. The sound of speed.

In a blink of an eye, a rider appears over the brow of the hill. The machine's engine song follows it up the incline, trying to keep up with the bike itself. As it hits the crest, the rider is already backing off the throttle. The front wheel is off the ground, and the bike is moving well over 120mph. The balancing act begins.

The rider moves precariously close to the left hand

Future World Champion Kevin Schwantz on the Team Lucky Strike Suzuki RGV500 in 1991. The factory works with the rider, exploiting that individual's strengths and designing around areas of weakness. Change the rider and most likely the performance of the team will suffer. In Grand Prix bike racing, the rider is the bike—a partnership between man and machine.

Alex Barros on the Team Lucky Strike Suzuki RGV500 in 1993. The Grand Prix field at any given race meeting has more natural talent than in any other series anywhere in the world.

There is no amount of wealth that can buy a mediocre racer a competitive ride.

wall as he tops the ridge on the back wheel. He feathers back the throttle for the turn. Simultaneously, the front end drops and he sets up the bike for the upcoming corner. He sits up in the saddle, extends his right knee, and the engine note—finally catching up with itself—rises three, perhaps four, times as he downshifts.

The rider's eyes focus intently on what's coming up. The bike angles dead center at the corner. The carbon brakes glow red for an instant. Brakes off, the bike begins a lazy movement down the hill. Tires are in contact and sticking, and the rider's left knee scrapes the ground for what seems like an eternity, and suddenly the bike is upright, then flicked—almost slammed—the opposite direction for the following right-hander; a combination of skill and natural physics. From below it looks fluid. A simple transition from right to left to right again. The engine's song changes, and hard acceleration catapults the bike down the hill toward the pits and the pit straight.

Standing at the top of the hill at Laguna Seca during an international motorcycle Grand Prix is an incredible experience. Because up at the top of the hill—between the seventh corner and the Corkscrew—you can see a ballet of sorts like nowhere else at Laguna. How else can you describe it besides dance? It is a well-choreographed ballet.

Unlike watching the IndyCar race at the same spot later in the same month, you can actually see what this pilot is doing. When Michael Andretti brakes for the Corkscrew in his Lola IndyCar, his helmet might bounce and cock toward the turn, but that's it. That's all you'll see of the driver controlling the machine. But in Grand Prix bike racing, the rider is there in front of you, plying his trade. Exposed.

The beauty of Grand Prix motorcycle racing is occasionally punctuated by brutally violent crashes and career-ending mistakes that casual observers see as senseless. But to the enlightened observer it is just a

At the top of the hill at California's Laguna Seca Raceway, one of the fastest parts of the track. The machine's engine song is following it up the hill, trying to keep up with the bike itself. As it hits the rise, the rider is already bucking off the throttle.

Brakes off, the bike begins an almost lazy movement down the Corkscrew. All that's required is a constant throttle, and the machine will gain momentum willingly. Sometimes the rider controls the racetrack, other times the opposite is true.

flaw in the dance step. Grand Prix racing is a very unforgiving sport. In that vane, what sticks out more than anything else—more than the colorful international atmosphere of a Grand Prix, or the awesome machinery ridden by fearless men in racing leathers—is the talent out on the circuit at any given race. To a man, the Grand Prix ranks have more natural talent than any other series anywhere in the world.

This is not Grand Prix auto racing, where the top half-dozen drivers are the best in the world and are paid accordingly and the rest are, well, neither. There is no amount of wealth that can buy a mediocre child a competitive ride. The mount enhances the rider's abilities, and an average rider will do poorly indeed. These men are the best in the world, and what they do is nothing short of awe inspiring.

In 1989, I had filed my story for the following day and was walking the paddock, looking for news. I had already seen Dan Gurney and subsequently ran into IMSA sedan racing veterans Elliott Forbes Robinson and John Morton walking unnoticed through the crowds. I stopped the pair and, searching for a good quote, I asked them what appealed to them about the show.

Robinson pointed over his shoulder at the sharp turn-eleven left-hander and said with admiration and wonder, "You know, I watched Kevin Schwantz come through there yesterday. I watched him get the rear end sliding. And I watched him feather off the throttle and steer the bike with the rear wheel until it got trac-

tion again. He set the front end down and never lost a position." He paused for a second and looked at me, then back at the turn. "Now that's racing," he said, eyes glowing.

"It's hard to find the superlatives to describe what these guys are doing," Gurney said. "You try to find the words to describe it and you can't do it. These guys are the best at something that very few people in the world can do. If they can do it, they're all here. I really admire these guys."

Good praise coming from such an accomplished racer. And probably less than they deserve.

Grand Prix motorcycle racing is about ultra-talented men on machines that manage to tax even their skills. The bike is more like a bull; the riders like rodeo stars. But in this arena, there is no soft sod to land in. A bull gets up to, what, 10mph? Contrast that with Shinichi Itoh's record top speed of 202mph. By anyone's account, the men who ride these machines are special. So special that the machines really should be considered extensions of the racers' own bodies. Prodigy John Kocinski ate everyone on a 250, riding his own style on a bike prepared for him. When he went to the 500 team in 1991, the bike was more suited to teammate Wayne Rainey, who had the development time behind him. When Kocinski returned to 500s a second time, after a half season in 250s, Cagiva, which was far more accommodating of the young man from Little Rock, Arkansas, was able to give him a motorcy-

Laguna Seca's Corkscrew is like a set of rapids on a fast moving river. The aggressive riders will be through quickly and smoothly, taking advantage of the natural shape of the course. From below it looks fluid. A simple transition from right to left to right again.

The machine, like no other in racing, is an integrated piece of technology, where everything is inseparably linked. Alex Barros, known for sliding elbows on the pavement from time to time, takes his Suzuki RGV500 through the final corner at Laguna Seca.

cle that suited him. In five races in 1994 he won twice (then failed to win again all for the rest of 1994).

Is Kocinski on a great bike now? Was Eddie Lawson cheated out of greater success by Cagiva? Why was Doug Chandler unable to win a race on the Cagiva in 1994?

These provocative questions elicit many more questions than answers. The answers from the experts so far have been mixed, meaning there are no simple answers. The qualifying time at the 1993 Laguna Seca race was slower than the record set in 1990. The track hasn't changed, the riders still know the place, and the machines have more power. What explains this discrepancy? In a race car, if the driver fails, the team can see it over time. The engine produces a given amount of horsepower and, as monitored by telemetry or any other on-board recorder, the use of that power can be charted. If the chassis is the same as most of the others in the pack, then the poor performance can be traced back to the man in the driver's seat. Change drivers and you get an improvement.

In Grand Prix bike racing, all the above is true except that the rider is the bike. Manufacturers cannot blame the rider solely, for it is the factory which gives him his equipment. It is the factory which works with the rider, exploiting that individual's strengths and designing out the areas of rider weakness. Change the rider and most likely the performance will deteriorate. Grand Prix motorcycle racing, more than any other form of motor sport, is a partnership between man and machine. And the machine, like no other in racing, is an integrated piece of technology where every-

thing is inseparably linked. In other books in Motorbooks International's Anatomy and Development series, I was able to separate the components easily and segregate their duties. The parts made up the whole. But on a bike there doesn't seem to be individual pieces making up a whole, but a whole with adjustable components. For example, I had three sections in the original draft, one for the frame, one for the suspension, and one for the tires. Several times I tried combining all three, rationalizing that they were, in fact, the same part of the bike. The frame is the thing around which the bike is built, but change the suspension and you change the nature of the frame. If you must logically separate the components there would really only be two, the running gear and everything else. Tires, dampers, springs, swing arm, frame aerodynamics are all interrelated, as are engine, transmission, clutch, fuel, and exhaust.

More than in any other form of motor sport, Grand Prix motorcycle racing is an exercise in integration. Hopefully as you read this book and discover all the variables that make up a Grand Prix bike, you'll remember that the best bike in the world is only as good as the guy who sits atop it thinks it is. And in the final analysis, it is perhaps the integration and sense of confusion over the machine itself that has made Grand Prix racing so popular worldwide.

Opposite page, Elite Grand Prix riders make racing look like a well-choreographed dance rather than the chaos it can be. Here Eddie Lawson, wearing the number one plate after his World Championship, keeps an ill-handling Honda looking smooth.

CHAPTER 1 HISTORY

Hunched low over their machines, goggles fixed to their faces, the first motorcycle racers were a far cry from the men who compete today. Early motorcycle pilots raced not with one another as often as with operators of other forms of transportation, like horses, carriages, or bicycles. Quite often the motorcyclists lost.

The first organized motorcycle races—if that description actually applies—were conducted around 1895. Little more than crowds of motorcycle and automobile commuters who wished to go fast, the first contests were loud, lopsided in favor of the automobile owners, and dangerous. The bikes raced alongside the monstrous cars of the day, often being done in by the drivers of the huge beasts. The races started in one city and the idea was to reach the next city as quickly as possible. The secondary goal was to arrive there alive.

As with the sport of automobile racing, it was France which began to organize motorcycle competition. In 1904, the French created an International Cup Race series, which pitted motorcyclists on a closed course with strict rules. Although the effort was ultimately doomed because of an unrealistic set of regulations, the idea of a closed-course series was sound. To adjust the rules and implement punishment for those who disobeyed, the Federation Internationale des Clubs Motor Cyclistes was created. Eventually it would be shortened to what we recognize today, the Fédération Internationale Motocycliste. The rules inspired a structure within which motorcyclists could finally race with at least a modicum of safety.

The motorcycle itself, however, remained a basically hazardous machine. Like cars, four-stroke engine advances quickly increased speed. Unfortunately, little was being done with the frames or tires. What began as a bicycle frame and tires fitted with a lightweight, underpowered engine had soon metamorphosed into a bicycle frame and tires with a heavy, fairly potent powerplant. The twin-cylinder Buchet, for example, was a 2,340cc engine mounted in a single-downtube frame, the whole package weighing more than 350lb. The frame tubing had been enlarged during the engine's evolution to accommodate the motor's weight, but nothing was done to address the inherent handling ability of the machine itself. Shocks and springs had hardly been dreamed up, let alone adopted. The engine's massive girth and weight, coupled with the unsprung chassis made the Buchet incredibly difficult to ride.

The British, afraid of creating a class filled with bikes similar to the obscene French machines, created a series that was organized around lightweight, street-going machines, with the superfluous components stripped from the bike prior to the contest. The idea seems to have been well conceived, but, like autosport in Britain, it was ill-fated. Races as run on the continent (that is, races on closed street courses) would never fly in Britain. The ridiculously low road speed limits—4mph maximum—were to be enforced whether racing was organized or not.

Britain's distaste of motor racing came as a response to the near barbaric contests of the early 1900s where, like contemporary championship rallies, spectators lined the road for a glimpse of the incredible machines. Often as many as 3 million spectators turned out to watch, standing in the road, moving from their vantage point on the street surface only when the machines were within feet of striking them. Frequently it was too late. Missing spectators and avoiding cars which were doing the same became a major talent and survival skill for motorcyclists.

The alternative was to hold the races on a small island off the east coast of Britain, the Isle of Man. The first Isle of Man TT races took place on the island in 1907, and the race has continued—attached to world Championship status up to 1976—until this day.

With what amounted to a British ban on motor sport, proponents decided to fight back. If racing was

English racer Freddie Whitworth aboard his Douglas flat twin in the early 1920s. Note the mechanical disc brakes both front and rear. Mick Walker collection

Geoff Duke rode in the classic, smooth, tucked-in style preferred by racers before the arrival of Kenny Roberts' *hang-off style. Duke is pictured on the Gilera 500cc four cylinder at the 1956 Italian GP.* Mick Walker collection

not going to be accepted on the roads, an off-road alternative would have to be created. Hotel magnate Hugh Locke-King funded the project and Brooklands was born. Created as the first permanent circuit made of concrete, the track was instrumental in creating not only interest in racing but success for British motorcycle manufacturers. The residual success was that of the British rider and team. Brookland's first race was held April 20, 1908. The teams, mostly English, rode anything available. At that first race, makes such as Norton, Chater-Lea, NSU, Minerva, FN, Rex, and Peugeot were on hand, with the win going to a Peugeot.

By 1910, motorcycling as both a sport and a pastime was gaining speed in Britain and in the United States. American Jake de Rosier and his powerful chain-driven, 900cc Indian made headlines in the U.S. as well as in Britain and Europe. But it was on the European continent where the competition really was most fierce. And it was on the continent that the Grand Prix championship was beginning to take shape.

With a brief intermission for WWI, the first "Grand Prix" was held, appropriately, in France. The next year two more Grands Prix were added, one each in Belgium and Italy; then came a Swiss Grand Prix . . . and so on. By the late 1920s, there were six to eight regular sites on the Grand Prix circuit—though it was far less of a circuit than anything known today. The series was nothing more than a confederation of races; there was no championship and no incentive to contest more than one race a season, let alone the entire calendar. It was really a forum to showcase local talent, with a few regulars making rounds on the circuit.

Consequently, the format was basically similar to a car race: the events were three to four hour affairs that often failed to hold the attention of the crowd for that same period of time. The races were run with machines of several different classes in the same event, with a different winner for each class—run like the 24-hour endurance races we know today. Although the format was eventually shortened and the machines moved into separate classes with separate races for each, it would be some time before the championship issue was addressed. WWII broke out in the interim. At that time Germany, with the well-engineered, supercharged flat

Mike Hailwood aboard Honda's mighty 500cc four cylinder, Brands Hatch 1967. Honda mechanic, Nobby Clark, stands beside him. Hailwood's duels with Agostini and the MV Agusta are legendary. Mick Walker collection

twins of BMW and the two-strokes of DKW, dominated the field. Conceived by Nazi Germany's nationalist government to showcase Nazi supremacy, the machines—as well as Germany itself—was eventually banned from competition. This was fortuitous for Norton and other British motorcycle manufacturers, because the German machines were far superior.

In 1949, the FIM finally adopted the structure of a series, and the first Grand Prix World Championship—with only six events (six for the 500s, five for 350cc machines, four for 250s, and three for 125s)—launched a new era of racing. Several years later, auto racing would adopt the same format with the creation of the Formula One Grand Prix World Championship.

Prior to the series format, there had been a wide variety of motorcycles on the circuits. Though bikes of virtually every configuration had been tried, Norton,

with its relatively straightforward design, was dominant prior to and immediately after WWII. Norton captured seventeen World Championships and countless more Grand Prix wins (more than thirty at the Isle of Man alone). In the early 1950s, however, the multi-cylindered Italian machines, especially MV Agusta, began winning more and more races. Norton retired from racing in 1955, giving only limited support to a handful of riders until 1963.

Even before the ultimate demise of Norton in racing (and virtually the entire contingent of obsolete British machinery), British riders had begun a migration toward non-British bikes—albeit with some consternation. World Champion Les Graham moved to MV Agusta (Meccanica Verghera) in 1950, and Geoff Duke moved from Norton to Gilera in 1953. Up-and-coming British riders Fergus Anderson and Dickie Dale

Kel Carruthers was 250cc World Champion in 1969 on Bennelli's four cylinder. Mick Walker collection

MV's great rider, Giacomo Agostini, wheelies the 350cc MV-3 in the early 1970s at Cadwell Park International. MV was the dominant force in GP racing from the late-1950s through the early 1970s. Mick Walker collection

jumped on Moto Guzzis, forsaking their native-built bikes for more competitive machinery.

Moto Guzzi had made waves in the Grand Prix scene with a four-cylinder, water-cooled, in-line configuration set along the frame, rather than across it as the MV Agustas and Gileras had been. Guzzi's move was significant as it was quickly discovered that this layout cut through the wind more efficiently. This marked the beginning of the era of aerodynamics, and further efforts to reduce drag soon followed. Although the odd-looking "dustbin" fairings spawned by aerodynamic experimentation would eventually be banned, the parameters for Grand Prix racing were really established during this time. Design moved from the broad-stroke concepts of surpercharging and engine configuration to fine-tuning concepts and overall package concerns. It was amid this philosophy change that MV Agusta became the force in Grand Prix motorcycling.

MV Agusta's true dominance began in 1956. Carlo Ubbiali secured the 125 and 250cc championships for MV on the seemingly unstoppable lightweight Italian machines. Although topped in every individual category in 1957, MV Agusta, led by John Surtees, came back in 1958 to win all titles in all classes—an amazing feat that was repeated again in 1959 and 1960. In 1959, Surtees won each of the seven rounds of the championship, posting records at each track along the way. Surtees left Grand Prix motorcycle racing in 1961 to drive for Ferrari, eventually securing a spot in Grand Prix auto history books as the first and only man to post World Championship seasons in both cars and bikes. Much of MV's success has been credited to Surtees, who directed a great deal of the technical aspects of the MV's evolution.

The 1961 season proved troubling for MV Agusta, the team suffering more than just the loss of Surtees. A shrinking budget forced the company to downsize its racing program. It proved to be good timing, since they would most likely have lost the championship anyway.

MV's loss of the smaller displacement championships had actually started in 1946, out of a tiny warehouse built upon a Japanese bomb site. It was there that Soichiro Honda began an industry based upon a series of small, war surplus, two-stroke engines. Honda became the first Japanese motorcycle manufacturer to produce both the engine and frame. The company dominated at home, producing more than double the product of all the Japanese manufacturers combined. In 1954, Honda sought international exposure by setting its sights on World Championship racing.

Soichiro Honda visited the Isle Of Man in 1954. It was a trip destined to have an amazing impact on Honda and the Japanese motorcycle industry. Honda

Trading Two Wheels for Four

Only one man, John Surtees, has been truly successful in both two- and four-wheeled racing, winning the 500 Grand Prix World Championship three times (in 1958, 1959, and 1960 for MV Agusta) as well as the Formula One Grand Prix World Championship (in 1964, driving for Ferrari).

Many others have tried and have had some success in both motorsports arenas. Johnny Cecotto, came closest most recently after winning the 1975 350cc World Championship, then racing Formula One for Theodore and Toleman in 1983 and 1984, respectively. Cecotto failed in his bid for a Formula One World Championship, however (he remains very competitive in European Touring Car Championships). Cecotto was the last to even attempt matching Surtees' record. Surtees remains the only man to have reached the pinnacle in each sport.

Motorcycle racers are talented men—perhaps more so than the majority of auto racers at the same level of professionalism. So why, when the speeds are nearly identical, are there so many differences? Why can't bike racers become competitive on four wheels?

Eddie Lawson retired from Grand Prix motorcycle racing at the end of 1992, ostensibly to direct his attention toward auto racing. Lawson offers a possibility of a double sport champion—although he is perhaps too old to become a championship-winning Formula One driver. In 1993, Lawson turned his attention toward the IndyCar paddock with a ride in an Indy Lights car. Lawson has so far been able to translate two-wheeled racing into four wheeled racing, and at this writing looked to make a smooth transition to cars, even catching the eye of some IndyCar team owners. According to Lawson, there are just too many differences between the two to be able to bring any motorcycle racing skills into he cockpit with him.

"The two are just so different," explained Lawson. "I'm not trying to use any skills from the bike. I think what you see on the racetrack is very similar, but what you feel is almost opposite. On a bike there are so many things you can do. If the thing's not setup right or not working just right, you can change your body weight and you can apply more body english to adjust or compensate for the things that aren't right. Like, for exam-

To date John Surtees (left) is the only racer to have captured World Champion titles in motorsports' two premier CLASSES: 500cc Grand Prix (for MV) and Formula One (for Ferrari). He is speaking with bike builder Arturo Magni. Mick Walker collection

ple, if the springs are too soft in the front, maybe you can just keep your weight back or not be as hard on the front brake. There's just things you can do on the bike to not apply as much pressure on the front end.

"But in a car you're strapped in with a three-point harness and there's not much you're going to do with your body. So car setup is so critical. I've learned that. At Laguna Seca in one session we were second fastest and at the next, we're seventeenth—just because we had the wrong setup.

"It's kind of a give and take. On the bike, I've ridden it for so long that maybe I feel comfortable on it. I don't feel confined and strapped in. Where in a car, you feel like you're really confined, and there's no place to go. On the other side, with the car around you, you really feel a lot more safe. And of course, with four wheels you don't feel like you're going to high side it. On a bike, when the thing gets out of shape you say, Oh boy, I'm in trouble here. But when a car gets out of shape you really don't think so much about it. Maybe that changes on the ovals but on the road courses you don't really get that feeling of fear. I love driving a car. I really enjoy what I'm doing. It's anything but scary. None of that for me."

Eddie Lawson, shown here racing for Cagiva, has moved from GP bikes to Indy Lights.

One last four-stroke gasp from MV Agusta before the complete two-stroke dominance we know today. This is the final version of MV's 500cc four built for the 1976 season. Mick Walker collection

discovered that European bikes of the period were capable of producing two or three times more power than like-sized Hondas. The realization led to a focus on engineering—a philosophy that eventually spread to every Japanese motorcycle manufacturer and which still exists today.

Honda entered its first Isle of Man TT in 1959 and failed miserably. Honda engineers redoubled their efforts at copying and refining the traits of the best European machinery, and by 1961 Honda was able to sweep the 125 and 250cc championships. Its sales bolstered by the racing successes, Honda was building and finding homes for nearly 100,000 additional motorcycles per month. This did not go unnoticed by the other Japanese companies.

Where Honda had evolved quickly from army surplus two-strokes to a four-stroke engine design, Suzuki, originally a textile producer, took the reins from Honda and became the world's most prolific two-stroke engine producer. Suzuki's two-stroke expertise allowed them to win the first-ever 50cc championship in 1962. Armed with that engineering success, Suzuki repeated as 50cc champion in 1964. A for-ay into 125cc machines proved equally successful with a championship there in 1965.

Yamaha, whose organs and pianos were perfectly capable of ensuring the company's survival, diversified into motorcycle production solely out of interest in the machine and the possibilities for success which lay in its marketing. In 1964, Phil Read competing in the 250 class delivered Yamaha its first World Championship. Read would go on to win the 250 championship in 1968 and again in 1971. Kawasaki had also entered the fray, capturing its first championship—a 125 title—in 1969.

The 350 and 500cc championships, however, seemed out of reach to the Japanese. The four-stroke MV Agustas were still the machines to beat, and piloted by top-notch riders like Mike Hailwood, Giacomo Agostini, and Phil Read, the Italian company was able to stave off the Japanese threat from 1961 until 1974, posting fourteen straight 500cc championships.

By 1967, Honda had posted sixteen Championship seasons and nearly 140 Grand Prix wins. Along the way, it had raised the stakes in motorcycle racing to incredible levels, spending a seemingly endless supply of money. Stymied by MV Agusta in 1966 in the 500

The history of motorcycle racing tires really started with the man who changed the way the tires were used: Kenny Roberts. When Roberts applied flat-track riding style to the pavement, the tires bore the brunt of the punishment.

Eventually, tires changed as his relationship with Goodyear Tire flourished. Here Roberts rides the YZR500 for the last time (for show) at Laguna in 1984. Note the Dunlop sponsorship; Goodyear was gone from motorcycle road racing by 1984.

class, Honda announced its withdrawal from racing. Suzuki followed suit at the end of the year, and Yamaha reduced its racing efforts. The void left Kawasaki in striking distance of a championship, and they very nearly succeeded in capturing the 500 class in 1970.

By 1970 MV, still utilizing four-stroke technology, found itself suddenly under fire from the two-strokes. Led by Ginger Malloy's Kawasaki/Bultaco, Giacomo Agostini's MV was assailed all year long. By the end of the season, Agostini had held onto his number one plate by a scant twenty-eight points. Yamaha also mounted an attack, and by 1972, the company known for quality pianos had squeezed into second spot in the manufacturer's Championship.

Standing by the four-stroke engine to the end, Count Agusta, MV Agusta's patriarch, hired Phil Read to partner Agostini on a second 500 bike. Although the move would bring a championship in 1973 at the hands of Read, it served to alienate Agostini, who switched to Yamaha in 1974. Agostini exacted his ultimate revenge when he snatched the 500 title in 1975

from the seeming death grip of MV Agusta—the first 500cc championship for a Japanese manufacturer. Although Agostini was enticed back to MV in 1976, the glory days were gone, and MV Agusta withdrew at the end of the 1976 season.

By the start of the 1978 season, the Japanese invasion was complete. The Far East now dominated the grids, and the first thirty-three finishers in the championship that year—as well as the first ten spots at any Grand Prix race—were held by either a Yamaha or a Suzuki.

The 1978 season was significant for another reason. It was in that year that the embodiment of Grand Prix motorcycling made his debut—winning the 500cc championship on his first try. Yamaha rider Kenny Roberts opened a new chapter in Grand Prix racing with his participation in the World Championship.

When Honda returned to Grand Prix racing in 1982, the conversion to contemporary racing was complete. Americans had begun their dominance of the riding and Japan had become unstoppable as en-

Cagiva's 1984 square-four, technically a Grand Prix bike, was ridden by Jimmy Adamo in the AMA National series. "Cagiva started as a hobby," explained team boss Giacomo Agostini. "Mr. Castiglione started because he liked very

much to race." Though slow to get up to speed, by the early 1990s, Cagiva was deadly serious about Grand Prix racing, fielding such top riders as Eddie Lawson and John Kocinski.

gineers of racing motorcycles. With the exception of a handful of insignificant attempts by Morbidelli, Sanvenero, Chevallier, Elf, Fior, and Cagiva, the mechanical field was primarily Japanese.

Moreover, from 1978 to 1994—with brief interruptions in 1981 by Marco Lucchinelli, 1982 by Franco Uncini, 1987 by Wayne Gardner, and 1994 by Mick Doohan—every championship has been captured by an American. Led by Kenny Roberts from 1978 to 1980, the torch was passed to Freddie Spencer in 1983 and 1985, then went to Eddie Lawson in 1984, 1986, 1988, and 1989, to Wayne Rainey, who won from 1990 through 1992, and finally Kevin Schwantz who won in 1993.

If racing had been transformed by the Japanese and the Americans, the bikes themselves saw only relatively subtle changes. Not that changes didn't occur—in fact there have been some intense engineering enhancements, but the overall performance of the motorcycle has remained roughly similar. For the past twenty or so years, when automobile performance has

increased dramatically, a great deal of motorcycle performance has remained relatively static.

"The bike has not changed a lot," said Giacomo Agostini in 1993. "It's not like Formula One. We have had an evolution. But in Formula One where you had skirts and electronic things, motorcycles have been the same for many years. We have had an evolution with the power and we do have some electronics—like for example, the gearbox is electronic, and we have some electronic suspension. But most of the change has been in the power. We have a lot more power than before, but the thing is, the speed is not so much different than before.

"I remember when I raced at Daytona in 1974, I won there and my speed was 295kmh. Now it's 310." Then he added with some thought, "Formula One has nothing to do with the normal car. But Grand Prix motorcycles are very close to all the bikes."

Kel Carruthers, who won the 250 World Championship, and who now works an a tuner and consul-

Wayne Gardner's 1986 NSR500. The older bikes were bulkier and more sluggish than later models. Nevertheless,

Gardner finished second to Eddie Lawson in 1986 and won the Championship on the NSR in 1987.

tant for major 500 teams agrees. "Things have stayed the same . . . basically, for about ninety years," he laughs. "The biggest problem is that a lot of people look at Formula cars and try to put Formula car information into motorcycles. So far it hasn't worked. They've tried hub steering and link suspension front and rear. And it looks kind of cute and everything, but they always go back to the good, old-fashioned telescopic forks. The three big differences [between when I raced and now] is that the tracks were a lot different then they are now, the motorcycles are a bit different, and the financial side is a lot different—the sponsors are quite different.

"When I raced in Europe probably about 80 percent of the races were run on regular roads which meant the speeds were a lot faster than they are now. For the rider, the emphasis was on high-speed judgment and high-speed corners. The bikes didn't accelerate and they weren't as fast, but the top speed on the circuits probably wasn't much different than what it is now because the tracks were so much longer and faster. Quite frankly you didn't race around the slow

corners; it wasn't worth the risk and everything. There was no point in racing around a first gear corner, where the most you could pick up was a tenth of a second, when down the road in the next five miles there were all these 135mph sweepers where you could pick up maybe three or four seconds. The way the tracks are now these guys have to race around first, second, and third gear corners. And the racing is more—particularly on 500s—based around the fierce acceleration they get out of the corners. Okay, they've still got the fast parts, but it has changed. That only came about because the tracks changed and because the tires got better."

What has changed, indisputably, has been the professionalism of the sport. With Roberts came a dedication to winning and a seriousness to competition. Roberts, with his rider protests aimed at improving prize money and professionalism, forever changed the way the series was regarded. It is now a multinational entity with multimillion dollar rider contracts and huge sponsorship stipends. Kenny Roberts' talent notwithstanding, what makes him as historically significant as he is was his ability to adapt and change the way riders looked at circuits. It was a technique that was to change the look of Grand Prix motorcycle racing. Often accused of being difficult for the sake of it, Roberts' was not contrary at all, he simply saw the world differently than others.

"It was sort of nonchalant," Roberts said. "I had raced the San Francisco Cow Palace as a novice. And one of the things we did at the San Francisco Cow Palace to get your foot to slide along the concrete was to put a piece of Supertape on your boot. But the minute I touched my knee on the ground—because it's leather and soft leather at that—it started ripping the leather apart. I came in from practice and started putting Supertape on my leathers in practice. In my pit nobody at the time had ever even thought of it. Kel Carruthers, who was my crew chief at the time, wanted to know what the hell I was doing. And I said, 'Well, they're dragging the ground'. And he just went bananas. He said, 'You know, you're going to hurt yourself,' and he just went on and on and on. And I said, 'Well, look, I've got to have something . . . because when it hits the ground its jerking my leg because it wasn't sliding.'

"I really didn't think it was such a big deal. I just came in from practice and went and did it. That was one of the things that—I don't know—you'd have to sort of have that kind of experience . . . to come from the Cow Palace and know that you had to have that to allow your foot to slide with the bike. And to just come in and do that to my knees and to go back out; now, that was the standard for years. That was the standard up until probably 1979 or '80—nothing like they have now to make them slide. That was eight years later. It was about '71 or '72 when I started to drag my knee."

"The last year I rode was '73," said Carruthers. "Seventy-three was the first year we had slicks. Kenny and I were really the first ones to test Goodyear slicks. And once they developed the tires, it made sense to ride the way Kenny rode because you could then lean the bike over and quite frankly you could get your knees to touch the ground. From then on, frankly, it revolved around the tires. The tires got better and better and better. Everything else improved, like the chassis and the engine and everything."

Eventually the style Roberts was still developing took control of the motorcycles. Roberts' says the best feature of a modern bike, compared to a twenty-year-old Yamaha is the stiffness of the chassis and the extent to which is has evolved. Roberts in large part accelerated that development by placing previously undreamed of demands on the chassis.

Roberts' technique was to come into the corner wide as if he were about to run off the road, then drop the bike on its side at what was considered a horrify-

"I remember when I raced at Daytona in 1974," Cagiva Team Manager and former World Champion Giacomo Agostini said. "I won there and my speed was 295kph. Now it's 310. So after maybe twenty years, the speed had not changed." Agostini remains the most successful rider of all time with fifteen World Championships to his credit.

ingly acute angle and accelerate hard out of the turn, allowing the rear end to slip and slide itself until the machine was pointed the right direction. Again, it was just an idea he tried on a lark.

"Sliding the rear wheel came basically from dirt racing," he explained. "That was my style as a dirt racer. I was always steering with the rear wheel. There are still a lot of people that don't steer with the rear tire—in fact, more people probably *don't* do it than do. And that's just the style, and when you have a style like that it carries over into road racing. You don't change your style of riding from a dirt tracker to a road racer. Sliding the motorcycle was something that I did in dirt racing that I carried over into road racing. It's almost impossible to do stuff like that without dragging your knee."

Roberts began another trend besides sliding the motorcycle and knee dragging. He started a process of intellectualizing racing.

"When it all came about it was from thinking. Ontario, California, was always the last race of the year. That was where I learned to use my mind to race. There was one corner there that was a U-turn, almost a 180. There was one point in the corner, the apex, which was in the exact middle of the corner, where I always felt like I was going to crash. The last four or five laps of my last race I thought, You know what I'm going to do?—I just thought about it because I was racing by myself in about third or fourth

In 1984, defending World Champion Freddie Spencer rode the first NSR500 at Laguna Seca in a non-championship event. Spencer won the 1983 title on a V-3, but here rides a V-4. This bike had the exhaust routed above the engine through what looked like the fuel tank; the tank was actually located below the engine.

place—I'm going to lean off the bike here. The instant I did that I said to myself, That felt decent. Then the next lap it felt better. They had two races back then. When the race was stopped, you put new tires on, and refueled, and whatever because there were two heats. And I sat down and thought about that, and I went out in the next heat, and all of a sudden I'm racing ahead of Carruthers and people I wasn't racing ahead of before that event.

"All during the off-season, I thought about that. I never discussed it with anybody because I was on ground where nobody knew where I was going. And I came out the very next year—I had an accident the week before Daytona so I couldn't lean off the bike, and I felt uncomfortable at Daytona—but the very next race when I was able to, boom!, I was leaning off the bike and dragging my knee.

"When I went to England the following season and raced on racetracks I had never seen before on the big 750s, I was racing against guys who were World Champions—or capable of being World Champions—the best Britain had, and Britain, at that time. was the leader in motorcycle racing. You had very little practice and you had to be able to put together the racetrack—had to put together where you were going to be on the racetrack and what gear you were going to be in. I was able to run through all that [mentally], but to do that I had to sit in the truck behind some tires where nobody could bother me. Well, that wasn't the kind of practice the English were used to. So right away, they were saying I wouldn't talk to the press, they said I was unapproachable and all these other things. One English guy figured it out. He asked permission to talk to me and I told him one time, well, I can't talk to you

Kenny Roberts' YZR500. Kenny and the bike were tire testing when this shot was taken.

right now, I'm right in the middle of a lap," Roberts laughed. "He couldn't understand that. But he was the only one who eventually did. And that was when everything started to happen, because I'd come out of that thing and I'd come out of the track with a lap record. Those kinds of things just kind of came to me either thinking about it, or naturally, or at the Cow Palace, or from dirt racing, or whatever, and I just carried it on to road racing, whereas people before that didn't have the opportunity to discover that stuff. I'm sure it would have happened."

The bikes improved to compensate for Roberts' style. In order for the rider to lean the machine over as far as Roberts did, the frame had to be strengthened. At the same time, the tires, which had already been redesigned, were about to go through another metamorphosis. Roberts' style invaded Grand Prix racing. And it dominated the Grand Prix scene for more than fifteen years. The changes he inspired have helped the overall performance of motorcycles. Broader powerbands, more predictable tires, and better chassis are Roberts' legacy. But only so much can change. Even

Roberts couldn't dictate a retailoring of the bike. After the revolution in riding style, the machine remains relatively similar.

"The bike doesn't look a lot different," said Kel Carruthers. "I think that's the good thing about motorcycles. They don't change a lot. The good things about motorcycles are they don't need a lot of rules for them, unlike Formula One cars or something. For Formula One cars they have to rewrite the rules every year. With a motorcycle the fact that you have to lean it to go around corners, for example, [decreases your options]. You can't use ground effects because the bike isn't big enough. You can't make the wheelbase longer so it doesn't do wheelies otherwise it can't go around corners. The motorcycle is pretty much self-governing the rules. Consequently, their appearance doesn't change all that much. The tires have gotten bigger, and the brakes have improved. Because of the tires, the chassis and forks are lot stiffer, but its just logical development."

Agostini now uses his past experience in motorcycle racing at Cagiva, which has a propensity to try

Fast Freddie Spencer. The only man to have won both the 250cc and 500cc World Championships in the same season.

Unfortunately, Spencer's career would soon wind down, even though he continued making attempts at repeating past glory.

new things. Helped by Ferrari, Cagiva experimented with a carbon fiber chassis and swingarm. Expensive development is rare in the sport.

"To change a lot we must invest a lot," says Agostini. "So it's important that things stay the same. It's not like Formula One where everybody must invest a lot of money. [People with less money] can't run a team because it costs too much money. So in motorcycles we like to keep it [inexpensive] to give the possibility to everybody to have a team."

"The problem, or the good thing, about a motorcycle . . . I think its a good thing," says Carruthers with some consternation, "is that it stops us from doing a lot of this stuff. A few years ago you had Formula One cars with 1,200hp and big wide tires . . . you didn't see them doing wheelies. They're long enough and the fact that they've got four wheels, they can use the force generated at one end of the car on the other end of the car. On a motorcycle you can't do that. A motorcycle with 45hp will do a wheelie. Consequently, you can't transfer any forces from one end of

the motorcycle to the other because it just won't do anything. If, for example, you try to take pressure off the front of the motorcycle when it goes down by hooking it to the back—all it's going to do is lift the back off the ground. There's no way that you can use a lever effect to transfer forces at one end of the motorcycle to the other like you can a car. It's like trying to pull yourself up by you're own bootstraps. It can't be done. Nowadays in Formula One its a little more complicated. But once upon a time they used to use torsion bars connected from the front to the back; when you put weight on the front you also put weight on the back. You just can't do that with a motorcycle."

"The bikes get better every year," Roberts comments. "No matter what happens, the bikes almost always get better. The 500s will run almost the same speed at three-quarter throttle as they will at full throttle; they will run almost the same speed at quarter throttle as they will at full throttle. They will go up their band at quarter throttle faster than anything anybody has ever ridden on the street. The 500s will leap out of

Amiable American Wayne Rainey won the World Championship three times and was challenging for a fourth title when his career was cut short by an accident.

your hands. [But] the bike will only go as fast as the front wheel will allow you to go. You can turn the power on a little bit more but the front wheel just comes off the ground. And there are very few places at Laguna Seca where you can grab a handful and just start shifting, even up the straight. We've got power cutoffs so you're not getting all the power. When I raced, the power cutoff that I had at Laguna Seca—coming out of turn eleven—was me standing on the back brake and shifting my brains out. You don't have to do that anymore. We have adjustments now that you can [use to] cut the horsepower down in the first two or three gears. It's adjustable. The bikes are much quicker than they used to be. They're not harder to ride—they're like a big kitty cat—but they just have so much acceleration now. Along with that, you've got to stop it. The thing will get over the hill like a rocket, but then you've got to stop it. That's one area that we haven't improved a great deal. I mean, the brakes are better than the bikes will stop. You can lock up the front wheel going over turn one at Laguna no problem. Except that you still can't get stopped, because

when you lock up the front brake you 're going to fall down anyway. So there are built-in problems in a motorcycle that you don't have in Formula One."

"The wheelbase that we use now, which is roughly 54 to 56in, is as long as it should be," Kel Carruthers said. "If you make the motorcycle three inches longer you just can't ride it around corners. We've moved things around so many times through the years. We know that you can't make it any lower than it is now because you can't make it go from side to side. It's going to get better. It's going to change. But it's hard to imagine it's going to be different. It still has to lean."

In addition to the technical aspects—or limitations—of design, there is also a matter of practicality. Unlike Formula One auto racing, where the sky is the limit, Grand Prix motorcycle racing seems to have a better grip on the cost of the sport. All three men, Agostini, Carruthers, and Roberts have changed professions within the field. Still at the cutting edge of the sport, they have shifted focus from riders to technicians, preferring to ride vicariously through their

Not all Grand Prix riders are alike, and different styles can still be successful. Wayne Gardner was able to ride his Honda even when it didn't want to be ridden. His uncanny ability to stay aboard the often ill-handling motorcycle won him a World Championship.

teams. "The more you win, the more pressure people put upon you, the more PR commitments you've got to do, and the more testing you've got to do because when you get that good no one can test for you," Roberts commented solemnly. "Pretty soon you build such a demand for your time and for your expertise that you just can't fulfill it. Then you've got other problems. I had a family, for example. I had three kids I had to raise. So I had to deal with that as well. All in all, it just got to a point where I said I don't want to do that. And all of the sudden it gets to a point where you don't have to anymore. You don't have little races that you can just come back and do. Once you've gotten to that plateau it's difficult to sort of back up four or five years because you've got a physical thing you've got to do. It's easy for a Formula One driver to come back to CART or sedans or whatever he wants to do

because there aren't really any physical changes. But laying off a motorcycle and then trying to get back on—no matter what kind of racing motorcycle it is—is very, very difficult to do. And its really the physical problems that I had at the end—muscles pumping up and just the lack of that type of training—that stopped me from doing the Laguna Secas and the Daytonas. Not that I couldn't race a motorcycle. I firmly believe that right now I can still ride a motorcycle competitively, but my body has to do it a lot. It has to be my job everyday and there was no way that I could make that my job every day like I used to."

So he turned his focus toward the bikes, the teams and the structure of the sport, as did Carruthers and Agostini. The Grand Prix World Championship has gone forward since the three retired—not to mention the distance it came while they were active racers.

CHAPTER 2 ENGINES

What's clear about a Grand Prix motorcycle racing engine is that is has incredible power . . . more than almost any other form of racing powerplant.

A quick street car—say an inline, 2.5-liter, dual overhead cam, twenty-four valve six-cylinder 1994 BMW 325i—has perhaps 190hp. The car weighs some 3,100lb and the power it develops is impressive, by all standards. By contrast, a 1994 Honda CBR900RR produces about 120hp and weighs 400lb—a power-to-weight ratio that leaves the BMW looking positively underwhelming. A Honda Grand Prix bike is just over half the size of the CBR900RR but produces nearly 50 percent more horsepower.

For the last several years, the choice of every GP team in the paddock has been a four-cylinder, water-cooled, two-stoke engine. Its amazing 180hp powers a bike with a total weight of 130kg—roughly 287lb. More precisely, that's approximately 1.5lb of motorcycle per horse. An IndyCar produces closer to fifty percent power to weight. That would be like giving the previously mentioned 325i 2,000hp. The Grand Prix bike has impressive power, indeed.

What is not so clear about the Grand Prix motorcycle racing engine is that it has *too much* power.

Racing engineering follows a traditional line of thought: If a little power is good, then a lot of power is better, and too much power should be just about right. But for years, the biggest problems in racing have not been producing more power—which they continue doing to the dismay of crew chiefs—but how to get it to the ground. Dragsters, for example, use multi-staged clutches to act as a buffer in the attempt to move the car from a standing start down the 1,320ft drag strip without spinning the tires. Fortunately, the application of power is fairly constant. That is, the track surface may slightly change consistency from track to track over the sixteen-race season, but essentially the same set of criteria is used to run the race: the track is flat, the car is starting from a dead stop, and the object is to mash the accelerator and ride the monster down the strip without hitting anything. Steering is used only for avoiding disaster.

The set of ingredients in a given motorcycle Grand Prix weekend are much different. If the motorcycle engine was analyzed for use in a similar set or corners for, say, an oval track championship, design preparation would be much more straightforward. Engineers would eventually find a way to get the power down to the ground.

Obviously, that is not the case. Not only do track layouts change throughout the championship, the various strategies in attacking the different circuits vary dramatically from rider to rider and from team to team. Do you sacrifice speed willingly in one corner to make it up on the entry of the next? What if your number-one rider happens to be able to ride through flat-out where no other rider can? The tracks are not flat, and the applications vary from corner to corner at any individual track.

According to the rules, engines in the premier 500cc class can have up to four cylinders. In 1994, the field consistently used a V-4-cylinder configuration with two cylinders up within the frame and two down toward the front wheel. Using a perfect horizontal plane for a reference, the rear cylinders are at something near a 45-degree angle. The front pair is at a negative forty-five degree angle, pointing toward the bike's radiator.

A run-of-the-mill, stock Yamaha YZR engine is cast aluminum, with nikasil-lined cylinders. The crankcases are currently cast aluminum, but were once made of cast magnesium. Cast magnesium proved too fragile for continuous use, however, with severe corrosion happening quickly and frequently. Pistons are forged aluminum, and for teams on a budget they could probably last two weekends, though the front runners generally change them at least once per weekend.

The fuel tank hides a great deal. The top set of pistons is accessible from above.

Metallurgy 101

Before we continue, a metallurgy lesson may prove useful. There are really three ways to create parts out of metal: forging, billeting or machining, and casting. Billeting, or machining, guarantees the most consistent results. The billet, or hunk or raw metal, can be perfectly manufactured with the desired properties ensured.

There are also three basic ways to change the structure of a metal. The first is obviously with the alloy itself. This is the simplest way to do it. It is also the most uniform. For example, a certain percentage of the alloy used in the manufacture of a normal racing block may contain a high percentage of silicon.

The second method is with the casting process itself. Cast metals will molecularly form one of three patterns: face-centered cubic, box-centered cubic, or hexagonal close pack. All this means is that the molecules will form different patterns. In the case of the first two cubic structures, iron (example only; no race bikes use cast iron) molecules bond themselves in boxes, where the hex structure is a honeycombed chain of molecules. In the box-centered structure there is an inherent design flaw in that the metal will tend to collapse under the wrong conditions (just like any square, concrete building).

Face-centered alloys tend to be more stable and will carry more weight, but they don't cycle heat well; box-centered alloys tend to be very stable in heat application. Hexagonal close packs tend to be more resilient as well as stronger, but typically flex more than the other two.

Thus by taking an alloy that is one type of molecular structure and mixing it with other compounds—like ceramic or beryllium (a high melting point, lightweight element often used as a moderator in nuclear reactors)—the metal's inherent molecular structure is changed. This change is called metal matrixing, which actually forms sets of bridges, or matrices, between the individual molecules. It fills in the spaces. More succinctly, it provides a bridge, or a frame, upon which the basic alloy can be strengthened.

The third way to change the characteristics of an alloy lies in the way the metal is tempered. Alloys can take on the same essential forms listed above if cooled properly. Now, if castings are the best way to form a complicated piece of metal, the heat it takes to melt it in preparation for casting will obviously change the makeup of the metal as it is being melted. A billet will have one property while cold, a second while molten, and a third as it finally cools after being cast. This is usually an undesirable situation.

But metallurgical engineers found that by quenching, or cooling the metal in different ways, the metal forms different end products—with a different hypereutectic state. Hypereutectic is the creation of a certain "grain" of metal, essentially frozen in a particular state by the way the metal is cooled, or quenched. For example, a horseshoe magnet made of cast iron has different qualities than a piece of iron staircase railing. In general, castings cooled slowly tend to be weak. The quicker they cool the stronger they are, but they also tend to handle heat poorly. Go to a foundry and ask what kind of metal they have. They will hand you a catalogue with fifty pages of descriptions of the many types of metal available! The point to all this is to emphasize that there are many different ways of changing a metal, from liquid-nitrogen quenching to cooling that takes several days.

Currently, the rules impose no limits as to what materials can be used within the engine. For example, the idea of using a ceramic coating has always appealed to certain designers. The goal here was to keep the heat from combustion from entering the surface of the cylinder head and piston crown. There has also been interest in the use of ceramic materials for the dual purposes of increasing strength and altering thermal expansion. With the secrecy inherent in the Grand Prix engineering domain, it may be some time before we know for sure if or how these material are used.

Two-Stroke Theory

The engine temperature must be kept as stable as possible otherwise the tenuous relationship between the oil, fuel, cylinder walls, and pistons becomes, well, blended. Literally. The point of the preceding metallurgical explanation was to give you an idea of what happens as the engine runs. Cooling liquid—by regulations, just pure water—is pumped from the bottom of the engine upward, and into a standard radiator which is exposed to air through the fairing. The engine is generally cooled to about 140-165 degrees Fahrenheit. Despite this, the guts of the engine are prone to failure.

The oil is spread around the engine via the fuel mix, lubricating the crank and cylinder walls as it is in-

Honda rider Shinichi Itoh produced a top speed of 202mph at Germany's ultrafast Hockenheim track. Said Schwantz: "Itoh does a lot of the aerodynamic work for Honda, so he's probably going to be the slickest package out there because they build the Honda around him." The fact that the 1993 NSR was the first successfully fuel injected bike didn't hurt.

duced into the crankcase (more about this momentarily). The cylinder needs to expand at operating temperature, thereby keeping the rings sealed to produce optimum compression. In effect, the rings and pistons press themselves to the sides of the cylinder—the only thing keeping them from melting to the walls is the oil mix. The oil is supposed to stick to the walls and act as a buffer between the metal of the piston and the metal of the cylinder. It must always be there, and usually it is. On the occasions it is absent, the piston will immediately weld itself to the side of the cylinder wall. If the temperature is hot enough to destroy the properties of the oil, it will remove the buffer and allow the hot pieces of metal to cook together.

All Grand Prix engines are two-stroke. Why two-stroke? Most racing engines—in fact, all street car engines—are four-stroke, meaning that the piston travels

The fuel tank not only functions as a place to store the unleaded two-stroke fuel, it also serves as a decent billboard. Aero-style filler caps allow the rider to hug the tank without discomfort.

Plumbing for the water-cooled Honda NSR500 engine. These slits look haphazardly cut, but the flow of the air through the bike is very complicated, and designers have found that in the case of ventilation, form definitely follows function

up and down four times to complete a full cycle—one stroke each of intake, compression, power, and exhaust. The crank has to turn over twice for any one cylinder to complete a full cycle. Each of the four duties are necessary for the engine to produce power. And generally it makes sense to segregate them into four separate movements of the piston.

But two-stroke engines do not separate the cycle into four movements of the piston, but into two. Combining intake and compression with power and exhaust, the two-stroke engine is able to step up the power capacity of the engine. The reason is clear: if a single-cylinder, four-stroke engine is limited to say 15,000rpm, it is only making power one quarter of the time (remember it only fires once every two revolutions of the crankshaft or once every four piston strokes). That's an incredible waste of energy (which is why both auto manufacturers as well as F1 car builders are seriously looking at two-strokes as a way to further develop power). Thus the insistence that the engine fire twice as often, making more power.

"You'll never get the power out of a four-stroke," explained Bud Aksland, Kenny Roberts' tuner. "A four-stroke has a longer power stroke, but for power-to-weight and everything, you can't beat a two-stroke. Two strokes make probably somewhere in the neighborhood of thirty to forty percent more power for a given displacement."

Trying to make an engine that weighs perhaps 100lb is easier by getting rid of the valvetrain. The valve mechanism is extremely heavy. World Superbike engines are about 60 percent heavier than GP engines,

a fair amount of it attributable to the valvetrain.

So the four chores of the four-cycle engine are combined into just two strokes of a two-cycle's piston—intake and compression, power and exhaust—and are all done in half the time of a standard four-stroke street bike engine. Remember that these engines are turning over once every 0.020sec. Snap your fingers and realize that the Grand Prix two-stroke has in that time turned over some seventy times and the piston has been on the power stroke thirty-five times (if you were able to snap your fingers in 0.3sec).

Although a two-stroke engine is mechanically simpler and more efficient at producing power than a four-stroke engine, it is still a very complex design. By its very nature, a two-stroke is expelling burnt gas from its cylinder while almost simultaneously drawing fresh mixture in. The fresh mixture is drawn through a transfer port (it has been previously compressed in the crankcase). Timing is critical to all these events; none of them happen precisely at the same time nor are they neatly segregated. Some overlap of events is inevitable.

The Grand Prix engine has had to take on some technology of its own in order to facilitate this delicate balance between duties. Some changes are fairly sophisticated, others are fairly simple. First, the heads on a Grand Prix bike are little more than caps, like plugs on the end of a pipe. Essentially they perform the same function—they seal the engine. They perform a function based on their size and shape, changing what is called the "squish band." The squish band dictates how the mixture will compress and burn. For example, most teams currently use heads with the spark plugs

Even when applied at a constant rate, the throttle will always produce uneven response depending on the carburetors, power exhaust valve, and load on the engine. Computer monitoring show how much throttle the rider is using in a corner. The results show a higher percentage of efficiency now that more throttle is being applied with more efficient powerplants. That slight increase gives the rider that much more confidence.

centered directly in the head. By machining the head and piston to match one another, a pocket can be created that will squeeze the mix into the desired area of the combustion chamber as well as creating a nice, even, radially compressed mix (the outer edges compress the mix evenly and produce a uniform swirl into the dome). In a four-stroke, this process is more complicated due to the valving.

When the two-stroke was first developed, technology didn't exist that enabled a valve to open once every 0.020sec. Pneumatic valvetrain was not an option as it is today on the exotic F1 cars (and still isn't due to weight), nor did metallurgy exist like that which enables an IndyCar to rev to 16,000rpm. The reed, however, was available and is fairly simple. It sits flat on a surface which is exposed to the inside of the engine. The intake port has a reed that is mounted to a triangular block, at the end of which sits the carburetor. As vacuum is created in the engine, the reed is sucked away from the wall of the metal block, thereby allowing air and fuel into the engine. As the vacuum becomes a pressure, the reed is forced shut and the fuel/air is kept out.

Usually the reed valve is made of a piece of stainless steel or carbon fiber which is sucked and pushed with surprisingly precise movements. In the past, crank-driven disc valve set-ups allowed mixture into the crankcase only at predetermined intervals. The disc allocated air/fuel mixture intake as it turned in relation to the crank. The disc window allowed an intake of mixture into the crankcase through an appropriate crank angle change. The system was effective, but created both space problems (necessitating an area for the discs to be located) and delivery was imprecise as the disc tended to flex and wobble when it was exposed to heat. Seldom were they true for long periods of time. Reeds, despite their simplicity, are more precise.

The mufflers are designed to meet FIM standards. Noise has been a concern at the Grand Prix theater since the beginning of the sport. Decibel standards are set, and tough penalties are imposed on those who fail to comply. Here, carbon fiber is used to reduce weight and help dissipate heat.

The exhaust valve is more sophisticated. Called a power valve, it is actually driven by a small motor—like a computer disc drive's motor—that moves an eyelid type of mechanism up and down allowing the exhaust gas to move in and out with more or less restriction. The exhaust ports—Honda currently uses two, Yamaha uses three—are in the cylinder wall. A cylinder wall port exposes area faster than any combination of poppet valves that you could create in a cylinder head. The port is simply a hole that allows the exhaust gases out, and the eyelid theoretically allows the entire port to enlarge or contract. Top end power is enhanced by a larger opening. And to keep the fresh mixture from whistling out the exhaust hole, the gate was created as a way of conserving fuel.

The eyelid controls the height of the port and changes as a function of rpm, allowing more to be moved out as the powerband is reached. How the eyelid facilitates the movement of exhaust gas is based on the amount of exposure to the expansion chambers—or exhaust pipes. Remember that the exhaust stroke is nearly combined with the power stroke. Also remember that the intake stroke will be beginning as the exhaust gasses are leaving, at the bottom of the stroke. The fuel delivery, therefore, must be at the bottom of the stroke. Think about what the heads look like. The reason there is nothing in the head area but a small metal cap on the top of each cylinder is that there is no fuel intake or exhaust taking place at the top of the engine. Most of it takes place at the bottom end of the engine.

Again, the intake stroke is occurring as the engine is pushing exhaust gas out. The first obvious problem is that an engine with excess pressure sufficient to expel its spent fuel is not going to allow the fresh charge in, let alone provide a suction to bring it into the chamber more quickly. Enter the transfer ports. The transfer valves allow the natural pressure and suction of the engine to do the work of a fourstroke.

Unlike common four-stroke engines, the fuel mix does not enter the combustion chamber through the

Regulations stipulate that the farthest the motorcycle bodywork can extend is the end of the rear tire. The exhaust pipes are limited to the same restriction.

Water-cooled engines are the standard in Grand Prix racing. The radiator and some of the plumbing is visible on this Honda NSR.

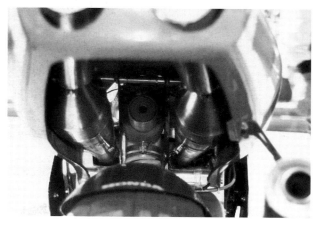

The exhaust itself is one of the most amazing aspects of a Grand Prix bike. Erv Kanemoto, Honda's legendary American tuner, has been seen at times with hundreds of sets of expansion chambers, looking for the perfect pipe.

Air enters the opening above the wheel and beneath the fairing to cool the engine via the radiator. The way air enters the fairing is not as important to most designers as how it exits the fairing. Designers are constantly trying to provide good cooling as well as good aero form.

valves above the piston dome (there are no valves anyway, remember). The intake is actually completed within the crankcase. As the piston moves up on the compression power stroke, it has already begun the process that will complete itself a full cycle later. It is pulling the fuel/air mix into the bottom end of the engine. The upward motion of the piston creates a vacuum in the *crankcase*. That vacuum is utilized to suck the fresh mix in. As the piston starts back down, it creates a pressure in the crankcase. The pressure forces the mix through the transfer ports—which are essentially sealed ducts within the water jacket and which flow around the outside of the cylinder toward the top of the cylinder. The mixture is flowing out of the transfers away from the exhaust port which is still open. As the piston heads up, it eventually closes the transfer followed closely by the closing of the exhaust port and then compression begins.

A GP bike's pipes are critical because the exhaust serves a dual purpose. First, as the engine fires, it releases the exhaust in a nice charge that travels out of the cylinder in a rush. As it travels down the pipe, it reaches an expanded piece that, as the sound waves expand outward, creates a negative pressure.

Originally, most two-stroke pistons had holes in the side of the piston itself. The fuel entered the crankcase through the side of the piston and the same vacuum and pressure through the transfer ports was used. Now, however, the intake is directly into the crank. Obviously, oil can be neither stored in the crankcase nor pumped into it via a dry sump. The way to oil the bottom end of the engine, then, is through the intake mixture itself—which is the reason the fuel is mixed with oil and lends that odd, sweet smell to the exhaust. The residual oil is eliminated in a cloud of smoke.

Engine Hardware

Enough two-stroke theory; let's consider the engine hardware. In the modern Grand Prix bike, each pair of cylinders works from a common crank and the two pairs—each pair on their own crank—are linked together on a planetary gear. Of the four major manufactures in 1994—Honda, Yamaha, Suzuki, and Cagiva—Honda is the only manufacturer to use a single-crank design. Not coincidentally, the Honda has been at the top of the power heap for years. Honda had reasoned that friction equaled inefficiency. To reduce fric-

tion meant to increase power relative to the rest of the dual-crank field. What resulted were eventually known as "Honda Lanes"—at fast racetracks the Hondas just pulled out and overtook with nothing more complicated than a twist of the throttle. However, the single-crank design has its flaws one of which is the gyroscopic effect it has on the motorcycle, making it difficult to turn.

1993 World Champion Kevin Schwantz felt it was the balance of the engine that made the Honda look as if it was always on the brink of disaster. "A lot of the times it's a little bit harder to ride than most of the other bikes. The Suzuki runs a twin crank that's geared together. What that does is reduce the gyroscopic effect. The Honda's single crank is spinning a lot more, which in turn will make it a little bit harder to turn and a little bit harder to change directions. Also, it's going to make it a little bit harder to stop because it's that much more mass spinning in just one direction. I think Suzuki and Yamaha looked at maybe giving away six or eight miles per hour at a real fast track and said, There's only one or two of those. The other twelve or thirteen races the Suzuki and the Yamaha are pretty much even. The Hon-

Rather than produce a high horsepower engine with a broad smooth powerband, engineers have produced a high-horsepower engine with a very narrow powerband. The droner makes all its power within 70 degrees or so of the 360 degree crank revolution. Thus, the power is unevenly distributed in a single crank revolution. Since a single revolution happens in about 0.020sec, the actual outcome is what feels like a broad base of power.

da might have a couple of miles per hour, but you can give somebody a couple miles per hour if you can get around the corners and through the chicanes quicker."

Again, too much power.

A problem specific to motorcycle racing is the struggle between power and handling. Front-wheel-drive and showroom-stock auto racers understand that to get through a corner faster often requires slowing the vehicle down, giving it more grip, and a better set for the corner. It seems appropriate for a motorcycle as well, yet sometimes it is not addressed that way.

Peak horsepower on the Honda 500 is about 180, and in the 250 it's near 100. But if the minimum power is a percentage of that, the power threshold would be half that of the 500—which means that the 250 will encounter many more situations where they can use full power than the 500. Energy is only available in packets. The motorcycle offers so much power that Doohan, or whoever rides it, must brake harder than the other riders. That shortens its "duty cycle" for the next corner. Honda increases the amount of power delivered during that time. In essence, Honda is

looking at a narrow, tall power pulse between corners, while Yamaha and Suzuki are looking at a wider pulse that starts sooner and is not as tall. Power has always been there; now the emphasis is on rideable power.

"More or less what's happened over the year is that the bikes have become a little bit less violent to ride," says Schwantz. "They are a little bit more predictable, I'd say. The Suzuki especially. This year [1993], not only was it a real predictable bike, but it was also real easy to get to work track to track. In the past, with the Suzuki, at some places it's been good, but trying to get it to swap from one track on one weekend to the next track the next weekend, we've had a real problem making the transition."

Single-crank problems notwithstanding, the Honda is much more powerful than the dual-crank engines. There are some fundamental differences between the engines. If Honda's single-crank design provides the right balance between friction and weight gains, the stroke/bore formula is responsible for the power that rockets Mick Doohan down the short straights between turns with jet-like propulsion. Some-

Unlike four-stroke engines, the fuel mix does not enter the combustion chamber through valves above the piston. The intake is actually completed within the crankcase and pressure forces the mix through the transfer ports—which are essentially sealed ducts within the water jacket and which flow around the outside of the cylinder—and to the top of the cylinder.

times the engine's output is overwhelming. Honda makes their engine at something around a 54x54mm bore and stroke, where Yamaha makes theirs 56x50. Engineers are still preoccupied by stroke because piston ring inertia is limited as stroke is abbreviated. Rings on a big-bore cylinder are badly affected by inertia forces, so at some high rate of acceleration and reciprocation—more than 100,000 feet per second, per second (gravity being 32 feet per second, per second)—the piston ring is inclined to jump off the bottom groove where it is held down by compression.

At the same time, using a nonexistent single-cylinder 500 as an example, the exhaust port is not big enough to move the spent fuel out quickly enough, so the port needs to be made larger. Eventually, in theory, the engine will lose ring performance, again due to the hot exhaust gas created from such a wide opening (the piston will also overheat). The advantage, often, lies in making the stroke longer and the bore smaller. The same is true with the four-cylinder. Once the engineer settles on the optimum port size the final job is getting the exhaust expelled.

The exhaust itself is one of the most amazing aspects of a Grand Prix bike. Erv Kanemoto, Honda's legendary American tuner, has at times been seen with hundreds of expansion chambers, looking for the perfect pipe. The pipes are so critical because the exhaust of a GP bike serves a dual purpose. As the engine fires, it releases the exhaust in a nice charge that travels out of the cylinder in a rush. As it travels down the pipe, it reaches an expanded piece that, as the sound waves expand outward, creates a negative pressure. Pulsating at 2,500ft/sec (mach 2.5), the pressure in the pipe is approximately 7psi below atmosphere. That creates a nice vacuum which, incredibly, draws almost all of the exhaust out of the cylinder. Obviously, engineers want the spent fuel evacuated as cleanly as possible with no leftover wasting room needed for the fresh charge. But if it continued straight out of the pipe it would inevitably produce a constant vacuum. Obviously, fuel needs to stay within the chamber at some time in order to provide the power for the next cycle. To facilitate this—and effectively shut off the vacuum effect—the exhaust is tapered into a small piece of

Moto Guzzi's Fantastic V-8

Through the early- and mid-1950s, Moto Guzzi had good success with their flat single in 350cc Grand Prix racing, winning championships from 1953 through 1957. It was compact, could be mounted low, and produced about 40hp—above average for the era.

But while toying with the idea of a 500cc entry, Moto Guzzi's Team Manager and technical guru, Giulio Carcano, realized a single would not be able to keep pace with the Gilera fours, which were producing in excess of 60hp. The Gileras, piloted by Geoff Duke, were far ahead of the field, and at nearly any venue on the World Championship tour, the competition was soundly trounced.

Carcano knew that the Guzzi team needed to change to a multi-cylinder engine to compete with Gilera. The question was, how many cylinders?

Because Carcano had spurned the design of the inline four, he could hardly afford to endorse it now as an alternative to his flat single. For reasons unknown, he also rejected a V-four concept. A twin-cylinder would not give him much better performance than a single, and a three-cylinder didn't seem to be the answer either. Unlike

Moto Guzzi's amazing 500cc V-8. Eight individual Dell'Orto carburetors nestled between the cylinder banks. The engine sat transverse in the frame. Mick Walker collection

current machinery which is limited to four cylinders, Carcano could design and build an engine with as many pistons as he liked.

In fact, he liked eight.

Carcano reasoned that if it were built right, the eight would not weigh substantially more than a four, would fit in the same space as a four, and would be more powerful—or at least as powerful. It was his reasoning that to be ahead of the competition, he had to think ahead of them, not just copy them.

Due to obvious space limitations, the engine had to be a vee. Clearly a straight eight would never fit within a reasonable chassis. The resonance from the long eight also would make for a very rough ride. A 90-degree V-8 was established.

Of course, it was a four-stroke, with intake handled by four cams and sixteen valves. Between the vee were eight tiny carburetors. An outstanding feature of the engine was that the swingarm pivoted from a mounting at its rear. Therefore the engine case was partially supported the rear end of the engine. This was done to help hold the racer's wheelbase to a reasonable length.

The engine was surprisingly light, weighing in at less than 100lb. In fact, the entire motorcycle weighed just under 220lb. It was also fairly compact, leaving plenty of room inside the dustbin fairing after it was installed.

The first outing was at Imola where it overheated ten laps into the race. At the next race at Assen, it took pole then blew up on the practice lap (a victim of a misselected gear). Finally, in Germany at the hands of 350 World Champ Bill Lomas, it raced with Geoff Duke's Gilera—at one point taking the lead—only to fail with a split water pipe.

Guzzi's V-8 would never win a race—although later that season it did post records for the standing kilometer (102-mph), standing mile (115.5mph), and the ten kilometer standing start (151mph).

The end of the era of both the V-8 as well as Moto Guzzi came at the end of that 1957 season, when Moto Guzzi withdrew from the championship.

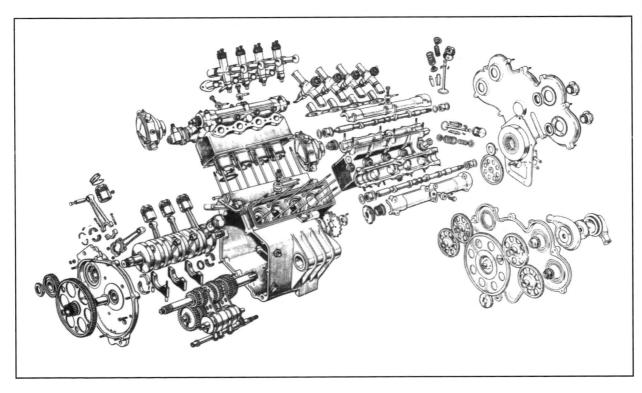

An exploded view of Moto Guzzi's surprisingly compact V-8. Mick Walker Collection

When the throttle is barely open, just a bit of mixture comes in. It is so severely diluted with residual exhaust, and the resulting mixture of exhaust product and fresh charge is so imperfectly mixed, that efficient ignition is quite unlikely. That's part of the reason the power band has such a strange threshold.

pipe again. The conical shaped reduction is something of an echo chamber. The sonic boom, created as the exhaust is released from the cylinder, is then blocked from moving out of the pipe. It literally bounces back and plugs the hole again. That pressure serves to keep the fresh charge in the engine. It is a combination of atmospheric pressure as well as sound waves which create either a suction or a blockage.

Intake is equally important and, again, very challenging to engineer and tune correctly. Although some bikes are fuel-injected, most Grand Prix bikes are carbureted. The basic idea behind carb adjustment is to minimize the peaks and valleys in the power curve. The air/fuel ratio of the intake mixture is critical because the demand for a particular ratio—based on a combination of factors—is constantly changing. The more often you supply the cylinder with just the right mixture, the fewer peaks and valleys you will have in the power curve.

The four most important circuits in a carburetor are the idle circuit, the off-idle circuit, the needle circuit, and the main circuit. All of these overlap to some extent, the most extreme being the idle circuit which is still contributing even at full throttle. A carburetor functions due to the pressure differential across its jets. Neg-

ative pressure is created on the engine side, and positive atmospheric pressure exists on the other side. Because atmospheric pressure is ever changing, the jetting must change as well to maintain the ideal relationship between the two sides. Needless to say, electronic fuel injection, which makes changes on the fly, should be more efficient. However, the difficulty of getting a fuel-injected, metered charge into the crankcase and then up through the transfer ports in a fashion that riders are comfortable with has been a difficult task. But in 1993, Honda began experimenting with the technology. And at the ultra-fast Hockenheim track in Germany, Shinichi Itoh produced a top speed of 202mph.

"My dad actually got that on his radar gun," said Schwantz with a laugh. "I think it [fuel injection] is something that all the other types of motor sports are leaning toward. I think it's a more efficient way of fuel management. So why not try to work with it? It might make the bike use just a little less fuel and it might make it more efficient. I know a lot of the [two-stroke] engines are being fuel injected. I just got an outboard motor from Suzuki, and it's a lot better engine than anything I've had before that was carbureted. As far as street bikes go, I think fuel injection would be a big advantage. But for race bike use, maybe it would be,

What's clear about a Grand Prix motorcycle racing engine is that is has incredible power—more than almost any other form of racing powerplant. What is not so clear about a Grand Prix motorcycle racing engine is that often it has too much power.

maybe it wouldn't be [an advantage]. The Hondas have always been fast. Itoh does a lot of the aerodynamic work for Honda so he's probably going to be the slickest package out there, the most aerodynamic, because they built the Honda around him."

Mick Doohan, Itoh's teammate, had been offered the technology in 1993. He declined.

"At the moment I don't want it," Doohan said in 1993. "[Itoh] likes the bottom end, and he likes the mid-range [of the carbureted bike], but the top is not as strong as the [fuel-injected bike]. The races aren't won on top speed, the races are won on getting through the turns and in between each point. Top speed doesn't really come into it. If a bike gets a draft off another bike, its going to pass it anyway. It's really at the braking marks where one bike is faster than another. If you can't get it off the turn, you've lost the advantage. The other bike goes accelerating away and you've got to catch it. I prefer something that is a lot more rideable rather than something that gets a good readout on a speed gun."

So the debate rages on. "The carburetor's such a simple thing," says Kanemoto. "The problem with the fuel injection is that the majority of the factories test-

Two-Strokes Take Over

MV Agusta won more than 3,000 races, thirty-eight individual World Championships, and thirty-seven Manufacturers Championships—all on four-stroke machines.

Fittingly, the company to finally knock MV off the 500cc throne they held for seventeen years (1957 to 1974) was Yamaha, one of the pioneers of two-stroke technology. By 1975, Yamaha had secured its first World Championship in the 500cc class. The following season, there were almost no four-stroke machines to be seen.

In 1950, MV Agusta created their first four-stroke single, a single overhead cam, which evolved into the 1952 World Championship winner in the 125cc class. Although a larger, 250cc version of the engine was created to win the 250cc title in 1956, it was eventually updated to a twin by essentially connecting two 125s. The 250 won again each year from 1957 to 1960. The four-stroke design was seen as the standard bearer for MV.

Around 1957, Yamaha sought to enhance the performance of its 250 and did so by recreating a German Adler, upgrading and updating the German marque's by then fifteen-year-old design. Yamaha made an assault on FIM Grand Prix motorcycle racing in 1961, contending both the 125 and 250cc classes. In 1963, the team had its first 250 Grand Prix win at Spa in Belgium, and in 1964, with Phil Read as lead rider, its first 250 Championship. Read repeated in 1965 and 1968. In 1969, the first seven position in the 250cc Championship were held by Yamaha.

While Yamaha was winning in the 250 and 125 classes, MV was moving away from the smaller championships, focusing instead on the 350 and 500cc championships. If anything, MV had found that the smaller championships were distracting, embarrassing to lose, and that their forte was fast, four-stroke engines.

MV's 500cc engine, the MV500-4, was a four cylinder with double overhead cam and chain drive (originally shaft drive). The engine remained largely unchanged until 1966 when, assailed by Honda, the MV-3 appeared. The MV-3 had twin overhead cams and, by 1969, was fitted with four valves per cylinder. It was capable of almost 170mph in 1969.

Giacomo Agostini was winning easily in the 350 class on a 350cc MV four, so in the early-1970s, MV overbored a 350 to produce a 430cc machine which Phil Read rode to victory (later updated again to the full 500cc specs) which gave the 500-4 better than 80horsepower.

MV had proven in the smaller classes in the late-1950s, as they changed the design of a successful 250 in anticipation of more competitive 250 machinery, that they were not adverse to change. But with early failures in their two-stroke efforts, MV remained loyal to the four-stroke. With engines capable of running easily to 15,000rpm, valves were becoming a liability. Not only did they have a difficult time keeping up with piston travel (which led to designs such as Ducati's desmodromic valvetrain), but the weight savings of the two-stroke was obvious.

The attack on MV from Honda in 1966 and 1967 was met blow for blow by like machinery, but the match-up between MV and Suzuki and then Yamaha was not. If John Surtees had developed the MV's winning ways back in 1956, Giacomo Agostini was instrumental in continuing the streak longer than it had any right to continue.

Suzuki won their first 500cc two-stroke race at Ulster in August 1971 at the hands of Jack Findlay, and finished second in the championship. The following season, Yamaha won its first 500cc race at the final event at Barcelona. But Yamaha had

Suzuki's 250cc two-stroke square four. The engine was nicknamed "Whispering Death" because of its tendency to seize. Mick Walker collection

The MV Agusta 500cc four that John Surtees rode to a World Championship in 1956. Mick Walker collection

been shut out of every other race by Agostini's MV. In 1973, Phil Read moved to MV and somehow captured the championship, snatching it away from Agostini. Feeling betrayed, Agostini switched to Yamaha's two-stroke machinery in 1974, and gave Yamaha its first 500cc championship in 1975.

Agostini's triumph sealed the fate of the four-stroke. From that point on, every winning bike on the World Championship trail has been a two-stroke.

In 1977, Honda returned to the 500cc Championship (and back to racing in general) with the new four-stroke NR500. Honda reasoned that a four-stroke engine had to be able to rev twice as fast as a two-stroke to produce the same power. Twice as many revs for twice as many power strokes. Twenty-two thousand

rpm or so. Honda decided that it had the technical expertise to pull that trick off.

Their engine was blessed with the famous oval pistons and eight tiny valves, four intake and four exhaust. Each cylinder had a pair of connecting rods to keep the pistons from tilting and each cylinder had a pair of spark plugs. But the incredible revs conspired to keep the valves from ever being timed correctly and the overlap was severe. Although a degree of it could be designed into the engine, what could not be eliminated were the problems that the reduced clearance created, essentially reducing the compression ratio. The engine often seized and was difficult to control at different rev points. At the end of 1979, the idea was retired. No four-stroke Grand Prix racer has been seen since.

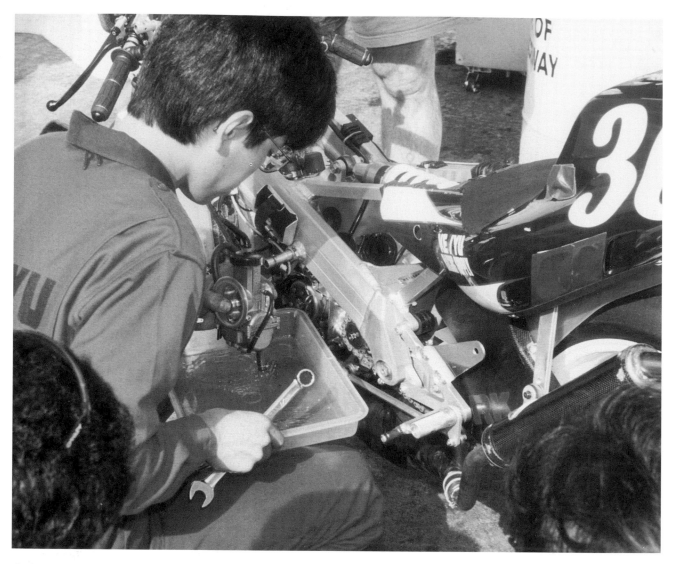

Carburetors are essential components and need to be tuned precisely. This 250 team was enduring endless problems with fuel delivery, frustrating them enough that they *disassembled the bike in pit lane after having wheeled it back to the garage a half-dozen times. It never did get sorted out during the session.*

ing it find that on the motorcycle it's not quite as simple as on the cars; they do run into some problems. Getting the fuel to meter correctly is one of the problems. Injecting the fuel into the crankcase obviously. It could be made to perform in a particular area, but getting it to perform over a wide range could be a particular problem.

"Let's say you wanted the thing to work really well through the range. You could do that, but normally it would mean a sacrifice in killing some of the power. You could get the thing to turn in the high rpm in an over-rev situation, but the programs that do that seem to present a problem in the lower rpm. It's hard to get it to work over a broad range. The first factory that can get it to do that I'm sure will have a tremendous advantage. I know for quite a few years there

were two or three factories that worked with it. It should be more fuel efficient, but in most cases it hasn't been, so we know in turn that it hasn't been working right. This new low lead fuel is going to be interesting, too, to see if it has some affect on engine performance. Possibly fuel injection can utilize this low lead fuel rule to its advantage. Right now the racing is really good, but I'm not sure where the future will be or where there will be a breakthrough."

The traditional method of improving lap times—the true gauge for performance in any road racing series—has been to go back to the drawing board to find more power. Power is the great equalizer. Or so the argument goes. Be it from fuel injection, expansion pipe tuning, or a low-friction single crank, the output is always being improved. But as we have previously dis-

Power will have to be scaled back unless the manutacturers think they can find some superhuman riders. Up-and-comer Loris Capirossi may be one of those riders.

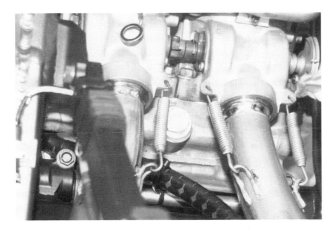

These top cylinders on the Roberts bike have a power valve that is linked between cylinders via a single shaft. Note the springs on the expansion chambers which allow easy access.

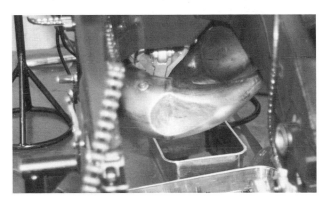

Some forms in the expansion chambers are for nothing more complicated than clearance. The chamber has to be altered to accommodate the tire and wheel.

The power valve restricts the exhaust at the beginning of the stroke. The rod actuates the valve and as the power comes on, it yawns open. The motor that runs the valve is similar to the drive on a computer hard disc.

cussed, power is counterproductive at a certain point.

From the 1991 USGP through the 1993 race, the lap times have actually gotten *slower*. Inevitably, the question must be answered: How much power can a motorcycle with it's limited tire contact patch handle?

If you watch any Formula One or NASCAR race on television, at some point during the two to three hour event, an in-car telemetry shot will flash throttle application on the screen. In any form of motor sport, the idea is to be full on power as often as possible. Flat-out power and hard braking. Full-on or full off. Obviously, in a sweeping corner the driver will need to feather in throttle, but for the most part, the driver wants to be flat-out as much as possible.

But in Grand Prix motorcycle racing the average lap is done at less than 10 percent full throttle. According to *Cycle World's* Technical Editor, Kevin Cameron, when the 500s raced at the U.S. Grand Prix

the riders were only able to use full throttle about *1 percent* of the time. That in itself is amazing. The engine, therefore, needs to make less horsepower.

"If you had stood in turn nine at Laguna where they come down the hill turning left under the bridge," Cameron said, "you would have seen 250s coming under there all solidly on the throttle, all making a big round sound, with the riders practically invisible from the outside of the turn because they were hanging off the machine on the other side, the bikes leaned all the way over. But in the same corner, the 500 engines are (almost idling) and the riders are not nearly as leaned over and are not going as fast. The reason is that they can't get on the throttle there because the minimum the engine can give is more than the motorcycle can tolerate."

Prior to the magical year of 1992, all changes came as a result of trying to increase power. Until, that

The aluminum cover allows access to the power valve for cleaning or adjusting.

is, the Grand Prix establishment decided to address how to better use what they already had.

What came along in '92 was, in fact, a better use of the power. In other forms of racing, these improvements had generally come from other places on the machine. In IndyCars it came from aerodynamics, in drag racing it came from clutches, in Formula One it came from traction control and semi-automatic gearboxes. In Grand Prix motorcycle racing, the changes came, oddly, from the engine itself. And it wasn't as simple as a mere reduction of power. What has been changed is the way the power is delivered.

What Honda developed in 1992 was a way to *apply* the power. That development—or refinement, since it was tried a dozen years earlier—has been called the "droner" engine. The droner, or "big bang" engine has been christened as such for the way it sounds—which is like a big, dull, unresponsive four-stroke engine. Or, more accurately, like a single-cylinder motor.

Says Bud Aksland, "In most racing—in most engines—you've got vertical twin and V-twins. With V-twins—like Harley-Davidsons—they usually fire two cylinders really close and then there's a big gap. That's why Harley-Davidsons were always superior on dirt tracks over Triumphs or Yamahas when you're trying to get traction. It's even more important on the dirt."

So Honda engineers translated the concept to two-stroke pavement racers. The concept is simple: rather than produce a high-horsepower engine with a broad, smooth powerband, the engineers have pro-

The YZR 500 is blessed with two cranks which aid in the balance of the engine. The cranks are coupled together via the clutch assembly which is bolted onto the end of the mainshaft.

duced a high-horsepower engine with a very narrow powerband. Instead of spreading the power load over a 360-degree range as a normal engine would, the droner makes all its power within 70 degrees or so of the 360-degree crank revolution. So the power is unevenly distributed in a single crank revolution. Since a single revolution happens in about 0.020sec, the actual outcome is what feels like a broad base of power. It

actually just supplies a violent amount of power in a short burst and then allows the tires to stick for a moment before supplying the next violent shot.

Confused? Let Randy Mamola explain how it feels on a Team Roberts' Yamaha: "The normal-firing engine tends to spin the rear wheel really easy, so you have to take it much more progressively. The first time I ever rode [the droner] was in Huntsville, Alabama,

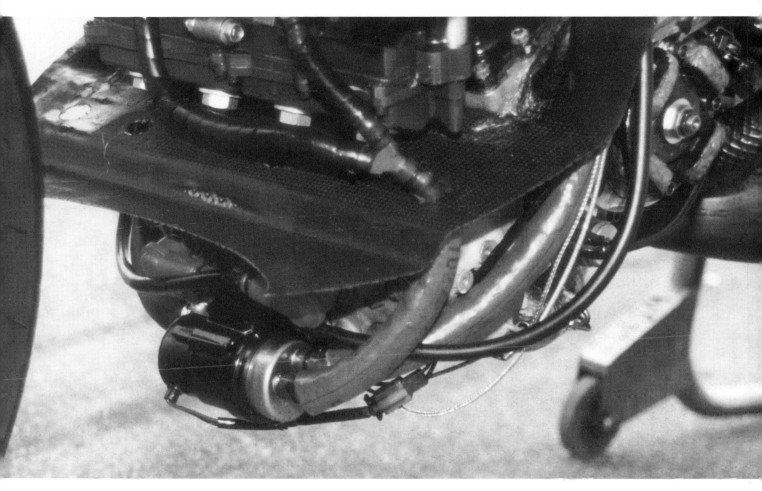

The fuel pump feeds all four cylinders from its position at the bottom of the engine.

where I tested rain tires for Dunlop. And even in the rain I was surprised at how much easier it was to ride. The way it fires is that it ends up catching the next cylinder much quicker than the standard style engine. Picture a wheel spinning, and it fires, then it stop, fires and stops, fires and stops—as opposed to fire, fire, fire, fire. These things put out 170–180hp and you've only got X amount of tire on the racetrack. So it's a great step forward as far as being able to ride the bike more comfortably."

Cameron calls it "ABS in reverse". It allows a tire, at the limit of adhesion, to get just a little break before giving it more power again. Once the tire is stuck, the next wave comes. Power has not been restricted, but redefined.

"It doesn't really knock the horsepower down," continues Mamola. "It doesn't feel like it's got less horsepower, it's just more rideable. If you can imagine an IndyCar trying to take off out of the pits and a street car doing the same thing. You have to be very precise to get the [IndyCar] car moving out of the pits. Same

with the standard two-stroke. You can get away with a lot more with the droner engine."

Computer readouts show how much throttle the rider is using in a corner. With these engines, a computer shows a higher percentage of efficiency now that more throttle being applied. That slight increase gives the rider that much more confidence.

"It's amazing how much easier it is to ride," continued Mamola. "It's very much like a four-stroke style engine. With the standard style motor they tend to want to spin easily. And in a standard style motor once you get them to spin, like I said, the engine wants to fire again much quicker so the tire will light up really easy. Like a V-8 has a lot of torque, and a smaller engine has a lot of revs. Basically it's the same sort of thing. So you're giving it more gas and the tire is spinning less so therefore you're coming out of the corner faster. Whereas, imagine it not being a droner engine, imagine it being a comparison between one bike with 200hp and one bike with 150hp. It's the same thing again. The bike with 150hp will allow you to give it

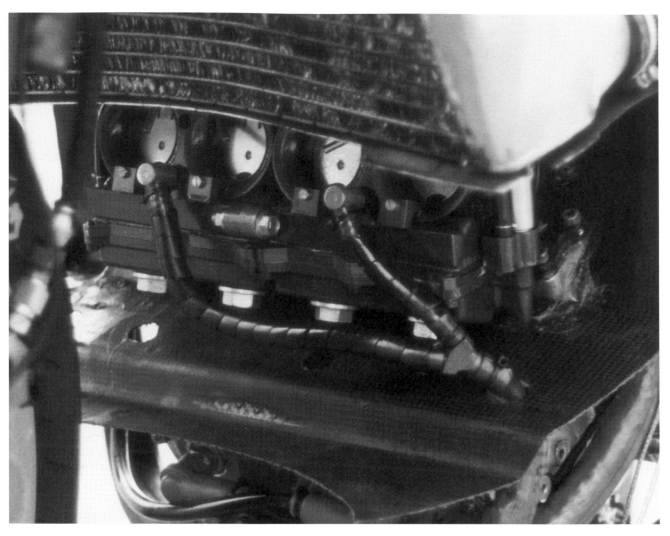

This Yamaha's bank of four carburetors is exposed to the air. In race trim, the air would not get to the carbs so easily.

more gas than the 200hp engine. But here, you have the same amount of horsepower—the droner and the regular two-stroke—but this accelerates a lot better because it's much more tame.

"When the engine pops at the one cylinder it puts out so much horsepower, then it pops again. Visualize that: one-two-three-four, then you've got to coast, one-two-three-four, coast. As opposed to bam, bam, bam, bam, bam, bam. You're adding the throttle so you've got one-two-three-four, stop, pause, then one-two-three-four, pause, one-two-three-four, pause. You can still spin it. It's very easy to do because you've got that much horsepower, but its much more controllable.

"Say we're spinning 10,000 to 12,200rpm. If you can imagine going in a straight line, our horsepower would start to reach at about 10,500 probably to 11,800, when it hits 10,500 and it gets close

to the maximum horsepower range, you wouldn't want it to not have the pause. Think about it: when it gets the power, when the power comes on, it goes from 10,000 to 12,000 like that." He snaps his fingers. "Therefore it would spin that much faster. Therefore you have a lot more gas coming out of the corner.

"A lot of times on the old-style Suzuki or Hondas I've ridden—Honda's three cylinder would have been the closest thing to the droner—say you've got a really long straightway, somewhere like a Daytona or something. You've really got to concentrate where your shift points are when you're getting up to fourth, fifth, sixth gear. At the end, when you're maxed out, the computer would check the to see where the rpm is. If you take the drone engine, because it's got so much more torque, it doesn't seem like it matters where you shift. Like when you drive your car: if your

Spark plugs are pulled after each session. These are extremely clean, suggesting everything is working well.

car revs to 6000rpm and you rev it to 6000 in second gear then go to third it wouldn't have the same pick up as if you shifted it at 3000. But with the drone engine, it doesn't seem to matter within 300rpm. It's got that much more torque."

The torque was initially a problem as it tended to break crankshafts and blow holes in crankcases. This was especially true for Honda, who was the only manufacturer running a single-crankshaft engine. With all the stress coming in just a few degrees on the crank, the testing was difficult and failures were severe. In most cases, the cranks' size was increased, along with an overall strengthening of the clutch, sprockets, and other reciprocating parts. The trade-offs for the new firing order were well calculated. For every part that needs strengthened, the overall performance suffers, so the compromise ended up being a seventy-degree power band rather than all four cylinders firing at once which was an option as well.

Future Changes

Sometimes the adoption of available technology is deliberately slowed to keep the costs of the sport from skyrocketing. Lessons from Grand Prix racing of the 1960s, when the Japanese factories spent truck-loads of money in the attempt to capture championships, as well as the huge expenditures in Formula One car racing have slapped the manufacturers back into reality. They move forward in some instances at a snail's pace. The new firing order was a fairly straightforward progression. Fuel injection seems to be one area where development would cost a great deal, and the path in that direction is not so clear. The small advantage may not be worth the cost.

"It's hard to make a big gain because the manpower and budget that you need to be a pioneer in the field is difficult," says Kanemoto. "The cost of running is nothing near Formula One, but it's hard to get that kind of a budget. If any of them [the manufacturers] ever got the budget of the Formula One car, you would see a lot more being tested. But I think right now everyone stays conservative because they can't put people on it. In some of the smaller factories they are somewhat committed to one idea—in the past there were some companies that worked on two or three engine configurations. Some of the other ones have to put it on paper only. Hopefully they are correct because they are pretty much stuck with the thing through the winter and through the year. But I know one company will go with two or three ideas and then narrow it down. It's all related to money I believe.

"They are nowhere near F1 standards," Kanemoto continued. "It's not that they are not able to do it. It's whoever may be the project leader or whoever. They've come a long way, but in the past it's been a matter of making the machine as simple as possible. All these people are afraid of some sort of reliability problem.

"I think you're going to see a lot more of that [advanced technology] soon, because HRC through the Formula One car racing team, has all that information and technology available to them. That's why they've been working with the fuel injection for the last so many years.

"I think there's constant development of the engine. It unlocks the thing to where you understand what's good and what's bad. It may be simply how to let the riders get on the throttle a little bit earlier. And it's not a simple thing anymore because everybody is obviously looking for peak power. They don't want to lose that on some of the faster racetracks, but it's a matter of being able to use the power. And some of it is the way the factories believe is the best balance. 1992 was an exception because Honda had a jump on everyone. But its kind of like plusses and minuses."

So as it has already, the focus will inevitably be redirected from solely power to usable power.

"Right now it's not a matter of who's got the most power, it's who can control the power the best," Kevin

This water pump is tiny, but its job is fairly simple. The liquid is cooled though a standard vaned radiator.

Schwantz explained. "Suzuki has an ignition and a lot of different things that help cut back the power in the he first two or three gears. A lot of times we'll run a small percentage of the power we could use in the first three gears until we get up into the higher gears where wheelspin is not such a problem. I don't think there's anyone out there—Honda, Yamaha, Cagiva—that's not restricted through the gearbox. Not necessarily by the gearbox, but gear to gear they are bound to be restricted somehow. Ours is done through ignition timing. It advances or retards the ignition timing to make the bike not pull as hard as it could in first or second gear. If you come out and you are less than 50 percent of the throttle, it might give you full power. If you're more than 50 percent in first gear, it will cut the power back 15 or 20 percent."

So as engineers continue striving forward with ways to create more power, remaining components will constantly struggle to keep pace. Inevitably, the power will have to be scaled back unless the manufacturers think they can find some superhuman riders . . . or a new application of technology.

"With traction control like they use on F1 cars now, I think we could benefit from another 25hp," says Kanemoto. "Without a little bit more advancement as far as technology goes, maybe not. Maybe another 10hp would be detrimental."

This Roberts engineer is peering inside the cylinders using a tiny flashlight to illuminate the piston tops and the cylinder walls. As high-tech as the onboard computers tend to make the project, one of the best ways to analyze the components is still with the well-trained eye.

CHAPTER 3 SUSPENSION

One Saturday afternoon in October 1992, Mario Andretti was introduced to a different view of Laguna Seca—this after having used the track for many years as a test facility and with ten years of races under his Nomex belt. On that day he was given the chance to ride one of Kenny Roberts' YZR500s. It opened his eyes.

Auto racers don't generally understand what happens on a bike in a corner. In a car, things are so

The dynamics of the rider, unlike an automobile driver, actually have an effect on the design of the bike. The machine can literally be built around the rider. A race car driver doesn't greatly change the attitude of the car. Perhaps a driver can be gentler or rougher with the brake and accelerator pedals, but in

the final analysis, the car stays upright, the wheelbase remains the same, and the amount of steering required to round a given corner for eighty laps without tearing up the tires is about the same for all other drivers.

simple. The car approaches turn eleven at Laguna in a straight line after having nearly 800 feet from the last right-hander to settle the chassis down. The corner is a 110-degree left-hander.

As a race car driver applies the brake, the front end loads up. The mass from the entire car concentrates itself on the front wheels. The nose dives slightly, and the rear wheels lighten up. As the driver turns the car to the left, the outside front wheel is being punished. Inertia is trying to keep the car going straight, and so the tire flexes.

As the car continues slowing, the balance will begin to reapportion itself. Each tire will begin to take up the load again. The driver will stab the accelerator and the car will move the weight to the back again. After the initial burst of power, it will have settled down, and

Opposite page, *the swingarm keeps the rear wheel solidly on the ground and acts as a buffer between the pavement and the incredible power the bike produces.*

as it passes the start/finish line, the weight will be equally distributed on all four wheels.

But on motorcycle, things are different—infinitely more complicated.

Like a car, the weight concentrates itself on the front during braking and on the rear under acceleration. But that's where the similarities end.

Once turn ten is completed, the rider will already be in the right place for the next corner. The bike will be vertical. On power for a brief instant, he will then begin to setup for the final corner. The rider will physically move back in the seat, sit up, and begin downshifting. Unless blessed with a mechanical or electronic device to stop it, the bike will dive visibly. The rear wheel may chatter as the front end takes all the weight of the machine. Like the car, the front is where the energy is concentrated.

"When you go into a corner when you're braking," Randy Mamola explains, "you're concentrating on what the front end's doing, because basically that's

Poorer teams don't have input in the overall design of the bike, and manufacturers don't build tailor-made bikes unless they're slated for factory teams. The small teams are allowed to make suggestions regarding adjustment and little more. Kevin Schwantz—seen here during the 1990 season—is one of the lucky factory-backed riders.

where all the weight is. You've got to start getting it into a corner. Now to get out of the corner—to hurry up and do that—you try to transfer the weight back on to the rear to make it steer."

The technique in bike racing has been, for the past fifteen seasons, to use a slight slip angle in order to help get the bike through the corner better. In a car, the relationship between rear and front tires is not really that significant. Basically it is a relatively stable rectangle. The wheels are balanced sideways as well as front to back. Needless to say, it doesn't fall over if left unattended.

But stand a bike up and look at it from the rear . The tires will be in line. That is, the rear tire contact patch will be in alignment with the front quite nicely. Sure it's a bit wider, but the contact patches line up well. Now lean the bike over at, say, a 45-degree angle. Take another look at the tire contact patches. If you draw a line between where the front tire meets the pavement and where the rear tire meets the ground you'll notice something significant. Not quite matching is it? Therein lies the first problem. The two wheels are rarely in line. Slipping the rear tire helps alleviate the problem.

"With a rim 6-1/4 inches in the back and with 3 to 3-1/2 inches in the front, you can picture the two wheels being out of line because you've got the two [greatly different widths] ," explains Mamola. "If you accelerate, the [rear] tire tends to collapse and it changes the profile, and it actually changes it so it steers better and better. The trail lines up better."

So the first problem a rider faces is the misaligned contact patches of the motorcycle. The stable platform the auto racer enjoys does not exist for the motorcycle racer. The rider must literally create a suspension at each corner from scratch.

But consider a couple of other factors: first, remember that there are gyroscopic forces in action. These forces keep the motorcycle upright. The spin-

Elf's Controlled Geometrical Variation (CGV) in action. The front swingarm was very difficult to get a real handle on *even by Serge Rossett (who directed the Elf program's more successful seasons).*

ning wheels and tires along with the crankshaft or crankshafts, have a gigantic effect on how the bike will handle. Leaning will create a completely different effect from steering with the handlebars. Essentially, a spinning object which is tilted will move differently than a spinning object whose direction is changed through movement on the horizontal axis. Because the bike relies on gyroscopic forces to remain upright, those forces are important. Not so in a car; when static it remains standing, tied together by four wheels.

In a car, as the auto racer applies the brakes it loads up the front wheels; the back end gets light, and the balance changes, but suspension geometry and wheelbase stay the same. Such is not the case on a bike.

As the rider applies the brakes, the front forks dampen the load, but they also shorten as well as flex to accommodate the weight. The rake and trail change as the forks move downward. Rake is measured at the top of the steering head and is the angle the steering axis is laid back from vertical. In effect, the movement of the forks change the steering geometry completely as well as shortening the wheelbase.

The effect of braking immediately alters the overall package. At one speed the bike has one rake and trail (trail is measured at the ground and is the distance

Mick Doohan on the Elf: "It just doesn't seem to work as well as [bikes with] telescoping forks because when you get on the brakes with telescopic forks the head angle changes to help you steer in. With a [front] swingarm, once you're into the turn it's okay, but it's getting into the turn that's difficult, because the geometry doesn't change on the motorcycle." Here Ron Haslam corners on an early Elf.

"You adjust the bike for certain racetracks," said Lawson, who has shifted his speed focus from bikes to cars since his Grand Prix retirement in 1993. "But you are limited. There are only thirty to forty minutes per session and only four sessions. So there are only so many things you can do. But if you went out testing, in a day you could get quite a bit accomplished. Once you've got the thing working, you probably wouldn't mess with it other than suspension settings." Nevertheless, suspension settings are critical to the success of a weekend outing.

Rider style has a lot to do with chassis design. Four styles of riding in one turn illustrate how much designers have to deal with when creating a chassis for a team. Even the two Lucky Strike riders shown—Schwantz and Barros—ride differently, posing both overall design inconsistencies as well as contrasts in suspension and chassis adjustment.

The bike is equipped with a pivot point that allows the package to bend in the middle. The swingarm holds the rear wheel to the ground and in order to do that it must pivot. No matter how hard he tries, a designer will not be able to design out the flex created by the hinge. It is not only moving, it is twisting constantly.

between a vertical axis bisecting the front axle and an axis extending through the center of the steering head) and at another speed it will change completely. Generally, a longer trail tends to make for a machine that is more stable in an upright position but takes more effort to go around a corner, whereas a shorter trail produces a quick-steering machine that may be prone to "head shaking" or sudden fits of uncontrolled steering movements.

For a motorcycle, the suspension is also the frame. That is, the movement of the frame must be considered in the overall suspension picture. It is, essentially, a large hinge. Comparing a bike to a car again, the car itself functions as a platform to which appendages are attached. The unsprung weight (wheels, tires, calipers, etc.) moves up and down at each corner individually. Although the wheels are connected by means of an anti-roll bar, they each really have autonomy. It is the ability of one side of the car to do something independently without affecting the entire car that makes a race car what it is.

But a bike is equipped with a pivot point that allows the package to bend in the middle. The swingarm

holds the rear wheel to the ground and in order to do that it must move up and down. No matter how hard he tries, a designer will not be able to design out the lateral flex that the hinge will create. So while the platform of a car stays stable—a given for that package—a bike is not only moving, it is also twisting constantly. In fact, many designers feel that—even during a time when the chassis has gotten incredibly stiff—that there can be no further improvement until the twist of the swingarm can be eliminated.

There is one final element that makes a motorcycle a machine with infinite adjustments: the rider's dynamics. Back to our car racer again. A driver doesn't greatly change the attitude of the car. Perhaps he can be gentler or rougher with the brake and accelerator pedals, but in the final analysis, the car still stays upright, the wheelbase remains constant, and the steering input required to round a corner for eighty consecutive laps without tearing up the tires is about the same for everyone.

But in the Grand Prix motorcycling ranks, the machine can literally be built around the rider. There is no right way to do it. Kenny Roberts found in 1978 that

Front Suspension Evolution

With a few rare exceptions—such as the exotic Elf motorcycle—telescopic forks have been the preferred method of suspending the front end of the motorcycle since the late-1940s. Front suspension has had the benefit of more years of use, compared to rear suspension, and more time for experimentation. For the most part, however, there are constraints which make the telescopic fork design the most reliable.

The front suspension is, of course, responsible for both steering input as well as damping. As with any suspension, it must also deal with the unsprung weight of the wheel, hub, and brake assembly.

Let us first focus on suspension as it relates to damping or absorbing bumps. Early motorcycles, like bicycles, were made with solid forks (or girders) that held the front wheel at its hub center and attached to the frame at the steering head. The front end was usually pushed back down to the ground via a single-rising-rate spring. The spring evolved also from a straight barrel to a torsion-based friction coil.

An inherent problem with the girders lay in their shape. Each piece was generally rectangular in shape, which was fine for suspension, but was completely lacking in side-to-side stability, tending to flex badly in cornering. The remedy was to make the girders round. Since they were going to be round anyway, it made

sense to integrate the round spring within each wheel. BMW did just that in 1935, patenting the oil-damped telescopic front fork, of which most motorcycles suspension has since been based.

Telescopic front forks were not without there shortcomings, however. There was lateral flexibility, and the hydraulic damping was far from precise. Grand Prix teams wanted greater rigidity and greater precision from their front suspension.

Two alternate styles were tried: Earles-type and leading front link. The Earles-type was an evolution of the telescopic forks in that on each side of the wheel a mounted spring pushed the wheel down onto the ground. The difference was that the Earles setup pivoted on a point behind the front axle. In other words, the wheel moved up and down as if it was on a vertical line based on the steering head adjustment, but it was held by a trailing link, which came off the hub at nearly a right angle. This design is typically associated with BMW motorcycles.

Leading-link suspension had forks as well, but unlike the Earles-type they were used for suspension mounting and wheel support. The fork was positioned at a positive offset (the forks were located far to the trailing side of the hub) and a set of springs and dampers were located just ahead of the forks and were connected by a

Earles-type fork on BMW 500cc Rennsport. Mick Walker collection

Above, leading-link fork on Geoff Duke's specially constructed Norton Manx. The large-diameter downtube doubles as an oil tank. Duke raced this bike in 1958 as a privateer. Such was his success on it that he was able to win factory rides from Bennelli and BMW the following years. Mick Walker collection

Right, a trailing-link fork on a 1951 NSU R54 500cc racer. Mick Walker collection

link from the solid forks to the hub. When the wheel hit a bump it moved up; attached to the solid forks by the links on either side, the links allowed it to do so, and the springs pushed it back down to earth.

Trailing link, another form, followed (no pun intended). Trailing link design was where the solid forks were mounted at a negative offset (ahead of the hub) and essentially operated backwards from the leading link suspension.

Hub steering was the final evolution, but never achieved the success that most thought it would. The system, which was first used in the early-1920s and which returned with the exotic Elf bike, is still a final point of departure for designers. Although it has captured the imagination of engineers from time to time, it has always proven ungainly and imprecise. The telescopic fork remains the standard for the foreseeable future.

Massive and rigid triple clamps mate the forks to the steering head.

to move a bike around the track the rider needed to slide its rear wheel. A few years earlier, a young Freddie Spencer found that he couldn't move his bike around a corner unless he shifted his entire body off the machine and hung off it like a child dangling off his father's sleeve. Remember that the gyroscopic forces change with respect to whether the bike is steered or angled around a corner. Roberts would tend to change the axis—relying slightly more on steering than leaning—and that determines the way it changes direction. One of Roberts' current [1994] riders, Luca Cadalora, has swung the pendulum of style back again to the pre-Roberts times when the rider and his angle of attack on the corner was more important than the ability to slide the wheel. Which is right? Who knows? But what is certain is that a motorcycle rider, as opposed to a car driver, has the capacity for incredible input into the machine. The rider alone knows what works on the bike and it is mostly accomplished by what feels comfortable.

The rider's attack on the corner is not only important in discussing racing theories, it is critical in assessing the dynamics of the bike in the corner. In turn eleven at Laguna, again, we put the package together: the rider brakes for the corner, the wheelbase shortens. At the same time, the front tire is loaded up, making the suspension sluggish from friction as well as increased spring compression. At the same time, the shorter trail makes the steering quicker. The rider eases off the brakes and begins to lean into the corner and the contact patches between the front and rear wheels are now offset. While the suspension is loaded, the bike doesn't just turn, it leans—or at any rate, the rider leans it. And it doesn't simply accelerate, it rockets. The weight redistributes itself as the bike settles into the corner and the rider applies the first bit of throttle. If the rider attacks the corner hard and late, he'll be spinning the rear tire to complete the corner. As it gets into its power band, the bike quickly gets power to the pavement, and the tire collapses. Depending on how the rider applies the throttle, the tire will change its profile. Once the bike straightens out and the tire is gripping better, the front end gets light and eventually will come off the ground. The rider is now essentially on a unicycle, and the steering will have returned to its normal settings.

The act of turning a corner is far more complicated than it seems. The idea of leaning a motorcycle

Left, *Grand Prix bikes have gone to inverted forks. Compared to many street bikes which have the sliders mounted to the axle and moving over the fork legs, the Grand Prix motorcycle has the slider mounted to the steering head and fork legs clumped to the axle.*

is based on a simple set of physics that deals with gravity, inertia, and centrifugal force. We all know that a pail of water can be swung upside down without ever spilling a drop—as long as it is swung quickly and with enough precision. Same with a motorcycle. Essentially, the bike is the pail and the rider is the water.

The motorcycle stays on the road, obviously, because of the tires. It is able to achieve its lean angle based on the grip of those tires combined with the force of gravity which tries to unseat the rider and send him flying. On any bike/rider combination there is a center of gravity. In every corner, depending on how quickly and how far off the rider moves his body, there is a formula as to how far the bike can lean. Like the water staying in the pail, the rider is simply fighting the forces of gravity via the tires' contact patches. The only

The forks serve two functions: they hold the wheel, allowing it to spin on its axle, as well as contain the dampers and springs.

As the wheel reacts to the road by moving upward, it compresses the fork. Inside the fork, the spring pushes the unit apart again, and the damping limits the bike's tendency to bounce. The fork works to maintain the tire's firm contact with the road.

variables between the tires and the forces of gravity exist in the suspension. The suspension can hinder or help the package by either allowing the tire patches to move within the overall package or by keeping them predictably beneath the package.

A combination of several things make up the suspension, and, in actuality, there is more to address than just the familiar forks, dampers, swingarms, etc. Suspension is also tires, frame, and steering. However, for this discussion we will limit ourselves to the suspension proper, with perhaps a foray into the frame here and there.

Starting up front, you have a telescopic fork set-up on every bike in the 500 GP ranks. The forks do two things: they hold the wheel, allowing it to spin on its axis, and they are the containers in which the dampers and springs are housed. Telescoping and at the longest setting while static, the forks will move inside themselves to absorb the affects of both braking and bumpy pavement—although certainly less of the latter

The fork is attached to the steering head via upper and lower triple clamps. Triple clamp design and material also affect overall front end rigidity.

The rear suspension consists of a swingarm and a single damper/spring unit. The rear damper does the same job as the front forks do, but as a single unit.

A potentiometer (the silver rod) is used to record rear damper movement. This data assists the tuners in establishing correct settings.

than the former. As the wheel reacts to the road by moving upward, it compresses the fork. Inside the fork, the damping action limits the bike's tendency to bounce while the spring pushes the unit apart again, reestablishing firm contact with the road.

Grand Prix bikes have inverted forks. Standard, non-inverted forks have the moving portions (the sliders) of the fork assembly attached to the axle, and these move over fixed cylinders (the fork legs) which are clamped to the steering head via the triple clamps. The inverted forks used on Grand Prix bikes are really just that: inverted. Essentially, the slider is mounted to the steering head, and the fork legs are attached to the front axle and move into the fixed unit. By having the sliders fixed to the steering head, their length could be increased. This is important because in either fork design the sliders contain the bushings that the fork tubes slide through. A longer slider means a greater distance between the bushing areas within the unit. This change results in less fork flex. A longer slider is not an option with a standard fork because it would severely limit suspension travel (the slider would smack the lower triple clamp).

Fork springs are chosen with the rider's style in mind. The rider can use either constant rate (the spring gives the same amount of resistance, increasing only as it is compressed) or rising rate springs (the spring gives one amount of resistance at one level and steps it up progressively as the spring is compressed).

The rear suspension consists of a swingarm and a single damper/spring unit. Unlike the front fork, where the entire assembly is hidden within the fork tubes, the rear suspension looks somewhat similar to that of a Formula One car. That is, there is a damper surrounded by a spring. The damper does the same job as the front forks do, but it does it as a single unit. As the motorcycle moves over bumps or accelerates, the damper and spring react to keep the bike from bouncing. But where the front forks are connected directly to the wheel, the rear damper unit is generally connected to a linkage of sorts which gives the swingarm more or less leverage and more or less damping, depending on what the teams are after. This is usually classified stuff, so trying to pin an engineer down as to what the strategy is on a rear swingarm is pretty difficult. Swingarm shapes vary in relation to the way the bike itself is designed. There have so far been few variations that have been shown effective at changing the handling characteristics of the bike to any extreme; the differences are dictated by such seemingly arbitrary factors as exhaust pipe shape. It is the hinge itself that is critical, and not so much the shape, according to one Honda crewmen.

As the swingarm moves up and down the linkages actuate the damper and spring, and the wheel is pushed up or down depending on the forces. The bottom of the damper is mounted to the swingarm—or its connecting linkage—and the top is mounted to the frame. The swingarm pivots up and down on a long rod or bold that goes from one side of the frame to the other. The frame will be discussed at greater detain in the next chapter.

In the case of both front and rear suspension, preload (the initial load applied to the chassis via the springs) is adjusted to oppose the weight of the chassis and rider. When the rider sits on the bike, the chassis will collapse some distance defined as "sag." Increasing pre-load reduces sag and vice-versa.

Cagiva saw the wisdom in being able to adjust the rear damping via the swingarm at a given race without having to go back to the machine shop. By using an adjustable pivot point, Eddie Lawson could come in from practice then return to the track at the next session with a very different handling motorcycle.

"You'd adjust the bike for certain racetracks," said Lawson, who has shifted his focus from bikes to cars since his 1992 retirement from Grand Prix racing. "But you are limited. There are only thirty to forty minutes per session and only four sessions. So there's only so many things you can do. But if you went out testing, in a day you could get quite a bit accomplished. Once you got the thing working, you probably wouldn't mess with it other than suspension settings. The first couple of years at Cagiva were pretty difficult because we had to basically alter the chassis quite a bit, and I think that they're done with that and I think they've finally got a geometry that they're happy with."

Lawson had the Ferrari factory helping with engineering the ability to change the design of the bike very quickly—which, as the story has it, is why the team became successful under Lawson. Apparently at Yamaha and Honda, Lawson's test rider abilities were not utilized to their full potential. The feedback the three-time World Champion gave the teams was squandered in the back rooms of the factories. Lawson wanted to see changes immediately, but they came slowly and were watered down and changed by committee when they eventually came. At Cagiva, Lawson's suggestions were immediately implemented.

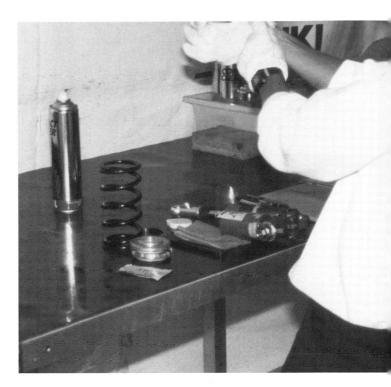

Rear springs are chosen with the rider's style in mind. The rider can use either constant-rate (the spring gives the same amount of resistance, increasing only as it is compressed) or rising-rate springs (the spring gives one amount of resistance at one level and steps it up proportionately as the spring is compressed).

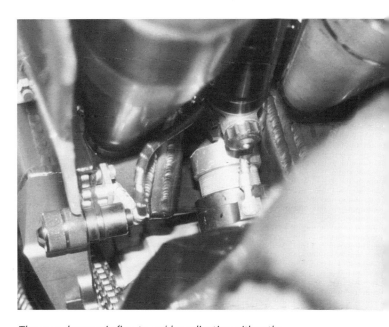

The rear damper is fine tuned by adjusting either the compression or rebound damping. The knob located on the shock unit is used to adjust the rebound and the remote knob on the left side of the frame adjusts the compression. The adjustments restrict the flow of hydraulic fluid in the damper cylinder.

There is very little actual steering done on a bike. The majority of the work is done with lean angles and weight shifting. Ten or fifteen degrees is maximum.

Other competitors don't generally have such impetus. Subject to the design of the moment they must rely on the adjustments they can do on existing chassis and suspension pieces, living with them until the engineers change the design back at the factory.

Geometry and Handling

Although not as significant as a full change in the chassis, there are several ways the bike can be changed to accommodate a rider's style or a particular track. "You could raise or lower the swingarm pivot," explained Lawson. "You could adjust the steering head angle or offset of the triple clamps. You have compression, rebound, spring pre-load, spring rate that you could change. But there are certain riders who like different things. Wayne Rainey liked the back of the

Teams can change the engine position; it can be moved up, down, backward, and forward. Teams need to determine if they want the weight over the front wheel or over the rear.

Do they want it higher or lower? By moving the swingarm, the bike's center of gravity—where the bike sits in relation to the wheels—changes.

Rear Suspension Evolution

Rear suspension had been more of a luxury early in motorcycling's history and a poor luxury at that. Most frames up into the 1930s had no rear suspension at all. The wheel simply mounted between a rear fork and was a solid part of the frame.

One of the early attempts at rear suspension was through a plunger-box style system where the wheel itself moved up and down within the mounting brackets. In other words, the point at which the hub was bolted to the frame was moveable. As the rear end hit a bump, the hub itself slid up and down within a box or tube. Each end of the hub spindle essentially was allowed vertical movement in one of two track-type components. The system worked poorly, as it required a great deal of chain slack to accommodate the straight-up-and-down movement. It also twisted badly.

Vincent offered the first truly efficient design in 1928. Its rear end utilized a triangulated swingarm similar in appearance to an unsprung frame, but pivoting at the point at which the bottom tubes connected to the seat tube. The top of the swingarm attached to two spring boxes located below the seat. The chain remained stressed and the wheel was free to move.

Most early pivoted forks were only slightly better than the plunger boxes they replaced. The forks were hardly strong enough and tended to flex badly. Different forms of triangulation were tried and manufacturers changed the width of the pivot to add torsional and directional strength. Of the early, simple pivoted forks, the Norton Featherbed was probably the best design. The Featherbed combined the pivoted fork rear end with a double loop cradle frame that had been welded in an innovative way at the steering head so as to lower the center of gravity. Later, various forms of triangulation and torsion sensitive metals (imbued with extra strength in one direction only) were both above and below the pivot point which aided rigidity.

Eventually, the double shocks were dropped, and a monoshock setup was adopted. After that, rocker arm assemblies appeared to actuate the single shock and coil spring combinations. Currently, all 500cc Grand Prix bikes use a single-spring, oil-damped unit.

Plunger rear suspension on a Norton factory racer, circa 1936. Note the swiss-cheese treatment of the plunger housing. Mick Walker collection

This detail of the Elf steering shows the relative complexity of the front suspension. What acts like a steering head bearing is mounted within the hub on this Elf X hub steering design. As impressive as it appears, its intrinsic flaws conspired against its actual performance.

bike real high with a real steep steering head and Kevin's is kicked way out; mine's in the middle. It's just rider preference."

As Lawson explained, the steering angle is extremely important to riders. The angle of the steering axis has an affect on how quickly the bike steers, thereby making it a common adjustment. The steeper the angle the quicker it will steer. The less offset the more stable the bike is and the less it wants to initiate a turn into the corner. Changing the offset of the triple clamps changes the trail. Obviously, all of these things interrelate, and the team can end up inadvertently changing several things at once with one adjustment.

Leaving the rake and trail adjusted to the rider's preference, a tuner can also move the steering axis forward or backward. This changes the weight bias which affects the center of gravity and changes the wheelbase.

A tighter racetrack puts a premium on quick steering. So on a tight track where a rider needs to go from left to right quickly, the tuners would raise the center of gravity and shorten the wheelbase for quick

steering. At a racetrack with fast, open turns where the bike is better off being stable and perhaps not steering so fast, the idea would be to increase rake and trail to lengthen the wheelbase.

Most tracks are a combination of tight and sweeping turns, so riders usually end up leaving the setting pretty much the same once they are comfortable with it. Tuners can then make the changes with a very small range of adjustments. The beginning of the season is where the big tuning comes into play.

Remember, there is very little actual steering done on a bike. The majority of the work is done with the way the bike is leaned and where the weight is being shifted. The steering is something like ten or fifteen degrees. The steering that is most often used is ten to twelve degrees maximum.

The rear end is suspended by the swingarm, which serves as both a means to keep the rear wheel solidly on the ground as well as a buffer between the pavement and the incredible power generated by the two-stroke engines. Although it can be considered as such, the swingarm is not really part of the chassis.

This photo, which shows the hub in relation to the entire bike, was taken before the Elf was raced. The steering hub was eventually re-engineered and appeared in a more simplistic—and perhaps more efficient—version than seen here.

"It works as a whole, but we refer to the chassis as only the main frame," explains Mike Webb, crew chief at Niall MacKenzie's World Championship Motorsports. "It's important that [the swingarm] is rigid and as light as it can be. It's also important that the swingarm be as stiff as possible. While its unsprung, it's on a swingarm pivot, so perhaps its not as critical as wheels, brakes, that sort of thing."

Unlike a car where the center of gravity is a major concern—and the goal is to always lower it—a motorcycle doesn't necessarily need the center of gravity low. In some ways auto designs are simpler because they are essentially a flat plane which engineers want to keep as low to the ground as possible. After all, cars simply turn corners. Because a motorcycle also leans when it turns, it has an extra direction of movement compared to a car. For that reason, the engineers might want to put some weight higher up on the machine so it will change direction more quickly. Tuners can alter weight distribution by adjusting the engine's location. It can be moved in four directions: up, down backward, and forward. Teams need to determine if they want the weight over the front wheel or over the rear. Do they want it higher or lower? By moving the swingarm the bike's center of gravity—where the bike sits in relation to the wheels—changes.

"At some racetracks it's very important to be able to change direction very, very quickly, so we'll raise the weight up," explains Webb. "At some racetracks, where perhaps we want it a bit more stable, we'll put the weight back down."

When Wayne Gardner and Mick Doohan were riding for Honda at the same time in 1992, there were a huge differences in their riding styles and, consequently, huge differences in how they liked the bikes setup. Gardner liked to brake late, turn into the corners quickly and exit hard. Doohan, by contrast, preferred to keep the corner speed up and drop the bike on its side further. Gardner liked the engine down in the frame because it tended to pitch forward less under hard braking, whereas Doohan preferred the engine high in the chassis because it allowed him better feel.

"Basically, it depends on the circumstance, what sort of problems we're having with the bike." said Doohan. "In a race situation, if you come across a situation where the conditions have changed, and perhaps the bike isn't working as well as it should, you just ride the bike the best you can from the feedback you're getting from it. There's not one position, there's a thousand different ways you can modify the bike for different circumstances.

"We try to set the suspension up to work the tires the best and as easy as possible, and also if it's going to slide to make it slide progressively. You don't want the thing too soft so that it's gripping and then firing you back out of the seat all the time. You want it at least firm enough that it will actually hold the slide so that when the tires do start to wear that it's controllable when the wheels are out of line. So it's kind of hard to point a finger at a problem you're trying to pick out. So you just try and adapt the suspension.

"You've been testing at that racetrack all weekend, you should have a half an idea what it will feel like in the race. You just try and set the bike up to kind of work. You know that after about ten laps the bike will be sliding around, so you kind of set the suspension up firm enough so that it will kind of hold itself up when the bike's out of line and when it does grip suddenly it's not going to squat in the rear and fire you back out of the seat."

As in any type of motorsport, racing is a compromise. Set it up for one corner and it may not be good for any of the others. It's always a tradeoff from one area to the other. But in this case, there are more variables to compromise in the overall package.

"That's the biggest thing to grasp is that its three dimensional and it does move in those three dimensions," Webb said. "Steering on a motorcycle has to do with not just steering, but in leaning it over as well. So the geometry of the front is certainly an interrelated thing that makes the motorcycle stable or not stable. There is a lot of compromising there so that you can turn a corner but not be so unstable that the bike can't take it."

"The suspension is getting more advanced each year," adds Doohan. "The tires are getting more advanced each year. When we kind of get into a corner, that's when we'll start throwing whatever we know at it, but as I say, each year we do test different engine positions, different steering angle positions, and we really haven't got anything that has given us better feedback than this chassis.

"It's sprung firmly, but not like a rock. It's progressive pretty much all the way through—it doesn't kind of get hard at the end or get soft. It's pretty much the same all the way through. It will hold itself up. If it hits a bump it will handle it pretty nicely rather than going rock-hard if you use a little more stroke on it."

There are many different approaches to making the bike handle, and many different ways to ride it—

all of which affect the suspension differently. The one thing that all riders can rely on is that once in trouble, the bike will react the same for almost everyone.

"Once the steering gets into the full lock position, there's only one way for the force to go and that's through the spring." explained Mick Doohan. "Once the spring is fully compressed there's only one way for it to go and that's back. And that's when a harder spring works against you because it flicks you out of the seat twice as hard. But in a normal circumstance [the suspension] will actually make the bike slide . . . the suspension's not absorbing the power as much as it should be—the tire is taking a lot of the power because the spring is keeping so much force on the rear tire that it's spinning the thing rather than absorbing it."

Up until 1994, most Formula One auto racing teams had used some form of active suspension (suspension that automatically and almost instantaneously raised or lowered the car as the sensors read the road and reacted to it). Similar technology had been tried in Grand Prix motorcycling years earlier—and is still being developed. But bikes, as stated time and again, are nothing like cars.

"The biggest problem," Kel Carruthers said, "is that the front end dips more than you want it to. There's not much you can do about it yet. There have been different systems. You can use hydraulic systems, where you use compression damping, or you can use a mechanical system that uses the braking force to hold the suspension up. The biggest problem is that then the forks are so sensitive. It's still got to work on the little bumps and when you try to hold the front up it stiffen the front end up too much."

So far, the systems have been run only one of those two ways. If hydraulic, it opens or closes a set of valves and increases the compression damping; if mechanical, there is usually a linkage system on which the brake calipers are mounted. The brake force pushes the forks up—or tries not to let it down, anyway. Still, the technology is in its infancy and as electronics are developed to handle the data that has been accumulated through the years, the problems may eventually be eliminated. As for now, the active suspension—or more accurately, electronically controlled anti-dive systems—are not significant to the performance of current Grand Prix bikes.

Like anything else, there have been new ideas. Many have worked, more have failed. "The tires and chassis work fairly well together as a combination at the moment," Doohan said. "I think if you put better

A telescopic front fork will dive as the brakes are applied. This potentiometer attached to a Roberts Yamaha YZR500 provides some feedback as to the movement and travel.

This box-section aluminum swingarm on Kevin Schwantz's Suzuki RGV500 is very similar to those used on other bikes throughout the paddock.

tires on it it would work better still. So I don't think it's outdated just yet. Hopefully it will be very similar for next year. I'm quite sure that's not one of the major things we have in mind to change next year."

"Within certain boundaries of what we know works," offers Mike Webb," there is nobody running anything wildly different; they are all basically the same. And within some boundaries, rider style makes up a lot. There are classic things like the fact that you setup different for a guy who rides really fast in a straight line, brakes really late, kind of squares the corner off, and shoots out, compared to a guy who leans it into a corner—you know, a classic motorcycle racer—and rides around the corner. You would set the bike up differently for those two different styles of riding. And anywhere in between there are variations."

As far as what is currently being tried, there are no radical designs, but in 1977, the French oil company, Elf, decided to enter the world of Grand Prix Motorcycle racing. André de Cortanze, who hailed from the automotive world, where he had headed developments for Ferrari, Alpine and Renault, and Peugeot. Using his background in the area of automobile chassis, his look at motorcycle technology led him to radically revise the concepts of the motorcycle's front and rear suspension assemblies. The key ideas were to preserve the constant attitude of the machine in both acceleration and braking configurations and to lower the center of gravity—which we now know was a mistake.

The first Elf-built motorcycle used a Yamaha 750 two-stroke-engine with the fuel tank located underneath the engine. The exhaust was located above the

The Elf front swingarm had virtually eliminated front end dive and the flex of the front end under deceleration. Despite the advantages offered by this style motorcycle front end, its drawbacks have kept it from being adopted by any of the major teams.

engine. What was truly remarkable, and remains so to this day was the frame . . . or lack thereof. The frame was nonexistent. Like an IndyCar or Formula One engine, the motor itself bore the weight of the machine. In other words, the frame did not hold the engine; the engine was, in fact, the frame itself.

To make the package more complicated than it already was, there was no traditional fork layout; instead the front axle assembly featured a "mechanically-welded, double superposed triangle, with a single disk at the wheel axis," as the press release explained it back then. A steering box mounted in the hub allowed the wheel to spin on the hub and move directionally within it.

After severe teething problems, the 1986 season saw the Elf complete its first Grand Prix season. The fuel tank below the engine was discarded, while the front axle assembly abandoned the superposed double triangle in favor of controlled geometrical variation

(C.G.V.)—which was just as strange (but slightly more adaptable). In 1988, after many patents and hundreds of different variations, Serge Rosset (who directed the more successful seasons of the program, and who will discuss chassis design in the following chapter) and the Elf quietly faded from the Grand Prix scene.

"It [hub steering] just doesn't seem to work as well as telescoping forks," remarked Doohan, "because when you get on the brakes with a telescopic fork, the head angle changes to help you steer in. With a swingarm, once you're into the turn—I believe . . . I've never ridden one—it's okay, but it's getting into the turn that is difficult because the geometry doesn't change on the motorcycle. On a motorcycle, the geometry is forever changing and you need that to actually make it rideable. That's one of the reasons why those configurations have been around for a while but they have never really been put to good use. They can never get them to work as well as a conventional set of forks. "

The Elf 3 departed from both then-standard bike technology as well as earlier Elf designs. Note the single spring and damper unit up front.

Although Kenny Roberts started the trend of powering the bike out of the turns back in 1978, the pendulum of style has begun to swing back again to pre-Roberts times when the rider and his angle of attack on the corner were more important than the ability to slide the wheel. Which is right?

Who knows. Luca Cadalora, seen here in 1994, has been criticized by his boss, Kenny Roberts, for not muscling the bike around corners as his predecessors did. Nonetheless, Cadalora won the USGP riding in a more traditional, tucked-in 250cc style.

CHAPTER 4 CHASSIS

In racing, everything is interrelated. Tire technology improves; suspension technology advances. But the ability to get the power to the ground is what all motor racing is about.

"It's getting the thing accelerating and braking as fast as you can," said Mike Webb, chief mechanic for Niall MacKenzie and World Championship Motorsport. "And when the tires and the suspension can cope with the power as the bike is leaned over, it puts more stress on the chassis and then it needs to be stiffer or needs to be able to turn quicker. So when the tires get better, that shows deficiencies in the suspension. And when [the suspension] can cope, it shows up the deficiencies in the chassis and you need to change that. And then you're back to tires again."

A never-ending evolution.

And from about 1988 to 1992, tires made huge strides forward. With the increased adhesion came increased loads on all the suspension components as well as the frame. So the components had to follow suit, undergoing a stiffening, tightening, and overall redesign process. Materials technology changed. Being able to make the chassis strong enough within the weight limit became paramount.

"Each year they go up in rigidity maybe twenty or thirty percent at a time," commented Eddie Lawson. "And they're finding that every time they increase it its better. But as you increase it you increase weight. So you have the trade off. Every year its the same old thing: 'We have to make the bike better but we have to stay at 130 kilos.' It's a tough situation to be in."

The frame is a fairly straightforward piece of the machine. It simply holds the package together. The standard way of constructing the frame is with two major sections, like a chicken wishbone. The U-shaped skeleton protects the guts of the bike and acts as the main support. Each frame spar is essentially one side of the motorcycle. They surround the engine and hold the hinged end of the rear swingarm. The fuel tank and fairings cover the frame, keeping it mostly invisible to curious eyes. The complete chassis cost for an independent team is about $150,000.

Again, the idea is to keep the frame as stiff as possible. Although at first glance the walls of the boxes look thick and heavy, with large brick-like sections, it is much lighter that it appears. And as you'll learn in a moment, there is more advantage gained by making it stiff than having it lighter but more flexible.

"On a weak frame, if it flexes one way, then usually what happens is that the frame will actually tie up and you put a big load on it," Lawson said. "Then of course, it unloads. Finally when the tire lets go, it will unwind. That's when you see these guys get into these things where it shakes its head or wobbles real bad."

So the idea is to produce a frame that will not buckle or bend under the stresses of acceleration. It follows then once the frame reaches maximum stiffness, changes need to be made in other places.

"You may have an advantage someplace in some of the corners and getting off some of the corners," said Lawson. "Let's say Honda is always best on the fastest racetracks. The engine influences the handling so much. Something that's smooth allows the rider to get on the throttle earlier. In turn, you're able to start the drive earlier. With some of the others, when you're off the throttle, waiting to get on the throttle you have a front end problem with the thing pushing and so forth. It's trying to make a balance with everything, so in turn, development is always going on.

"But sometimes what you believe will be good is not necessarily any better. Engine development goes on all the time, but it goes along with the chassis and everything. A lot of people will say, 'If they could only put that engine in with so-and-so's chassis.' That's not quite the case. The way the engine delivers the power will have a great deal to do with how the thing steers

Cagiva's chassis (this '86 500 appeared at the French Grand Prix raced by Juan Garriga) has come a long way since 1986. In the early 1990s, Cagiva experimented with an all carbon fiber chassis, then opted for just a carbon fiber swingarm, and in 1993, abandoned the material altogether. Most feel carbon fiber will be the material of choice for chassis builders by the 2000 season.

in going in the corner and how it steers coming back out of the corner."

As Lawson said, the frame has been getting stronger, stiffer, and heavier. But it is essential that the beefing-up occur, otherwise as the power continues its upward progression, the frame will make the bike unrideable.

So the frame holds the key to several puzzles of Grand Prix racing. It may be where the next revolution occurs. "I think," said Giacomo Agostini thoughtfully, "we will work more on the frame. The frame is very important. I don't know for the future, but I think the engineers will work a lot more on the frame in the future. Because you make the time in the corner. The speed is important but not very important.

In 1993, prior Wayne Rainey's career-ending crash, the title was between two Americans on factory bikes—Kevin Schwantz on his Suzuki RGV500 and

The frame is critical to the overall performance of the motorcycle. If the engine has a great deal of power, it's essential that the frame be stiff enough to accommodate it.

At the same time, if the frame is too heavy, the package as a whole suffers. As in every racing series, the package is more critical than any single component.

Rainey on the Yamaha YZR. Oddly enough, Rainey's machine was not built by the factory, but by ROC chassis. Roberts' Yamaha team had determined that a non-factory frame was the best way to go.

Racing Organisation Course (ROC) is a French company which focuses on two activities: car racing for French SuperTourisme championship and motorcycle chassis building. The reason the company is important to us is twofold: first, its president director

Serge Rosset takes care of the bike side of the business. Rosset, you will recall from the previous chapter, was the person in charge of development of the revolutionary Elf motorcycle. He was also the Team Manager for Freddie Spencer's Team France effort in 1993. Secondly, it shows that in going forward, sometimes teams can go backward—giving other manufacturers like ROC a shot at the championship.

"All my life," said the amiable Rosset in his thick,

A 1993 ROC frame, built by Racing Organisation Course, a French company which focuses on two activities: car racing for French SuperTourisme Championship and motorcycle chassis building. The Roberts' Marlboro Yamaha team opted to use this private chassis over that built by the factory.

When the tires and the suspension can cope with the power as the bike is leaned over, it puts more stress on the chassis. Then it needs to be stiffer. So when the tires get better, that shows deficiencies of the suspension. And when that can cope, it shows up the deficiencies in the chassis and you need to change that. And then you're back to tires again.

French accent, "since twenty-two years—I have made my own bike. I worked ten years for Kawasaki for endurance bike and we were several times World Champion. We have worked for six years with Elf for this particular project. We made forty patents and Honda used a lot of them on their bikes. Now I work for three

Tube frames long ago gave way to the modern Grand Prix frame which is built with what are called spars. Each spar is essentially one side of the motorcycle. Like the wishbone of a bird, the U-shaped skeleton protects the guts of the bike and acts as the main support. It surrounds the engine and provides the pivot point for the swingarm. The frame's relatively large sections look massive and heavy, but the walls of the boxes are relatively thin. The size of the section is the more important factor when it comes to rigidity.

or four years for Yamaha, and I have a contract with Yamaha to develop frames for customers. They deliver the engine, and we make the bike complete to have more accessibility to price, a cheaper machine to fill the grid. When we started this three years ago, sometimes on the grid there were only twelve machines. When we started the last race we had thirty-six machines. We are very pleased with that. These bikes for customers are now competitive. We have seen for example, Niall MacKenzie is seventh in the championship with the private bike—not a factory."

It could have very well been the World Champion, had Rainey not been injured with just three races remaining in the series.

The frame is critical to the other aspects of the motorcycle. If the engine has a great deal of power, it is essential that the frame be stiff enough to accommodate it. At the same time, too much weight and the package as a whole suffers. As in every racing series, the package as a whole is more critical than any single component.

"There's no upper weight limit," Webb said, "But you want it to be as light as you can, keeping it down pretty close to the minimum weight limit, yet making it strong. That's the major chassis concern. We worry about weight, absolutely. It has a great bearing because the whole motorcycle weighs only 130 kilograms. Any amount of unsprung weight is a large percentage of the whole package, so yeah, its important."

Chassis Evolution

A Grand Prix motorcycle frame is designed to do two jobs: first, it is expected to bridge the gaps between front and rear suspension and hold the engine and transmission in place, or in other words, incorporate all components into a neat and tidy package; second, it is expected to provide performance through predictable, precise handling.

Although experimentation with front suspension springing and damping was rampant in the early-1920s, it proved to be quite some time before any true advances were made in rear suspensions. In that respect, frame technology was limited. With no springing in the rear, the frame itself was relegated to simply incorporating the components.

The roots of motorcycling are linked with bicycling. Early designs were nothing more than engines bolted onto tubular, diamond-shaped bicycle frames. Examining a stripped-down frame from most any motorbike of the early-1930s would yield some standard design philosophies. The steering head brought together one end—both the top and front downtube(s)—of what was almost without fail a diamond frame. The engine was neatly and conveniently located within the space of the frame—essentially a pair of triangles joined by a common support—the seat tube. It was triangulated both to improve steering precision as well as take advantage of a triangle's inherent strength; a more-versatile rectangular frame tends to distort laterally under stress. There was usually a single seat tube, and the engine was typically secured at the bottom of the seat tube and the far end of the front downtube, the bicycle bottom bracket was essentially replaced by the motorbike's crankcase.

The bike frame concept evolved into a cradle which used two tubes running between the bottoms of the seat tube and front downtube(s) to support, or "cradle," the engine from below.

From there the variations became more subtle. The Cotton frame, for example, triangulated the frame more severely by utilizing long top tubes whose far ends comprised the rear wheel mounts. The engine served as a common mounting point to which both the seat tube and front downtubes were attached. Some manufacturers used similar triangulation, but omitted the seat tube entirely, instead mounting the seat on the double set of top tubes.

Other frame designs included beam types which were single pieces of metal which formed a kind of spine for the bike. Stressed member frames (using the engine as an integral part of the structural integrity of the frame) were often the focus of experimentation. Postwar Vincent V-twins, for example, were frameless save for a box-section "backbone" (which doubled as a dry sump oil receptacle) bolted to the top of the engine. One end of the backbone was connected to the steering head, and the other was connected to the swingarm via the spring and, later, damping units. At the back of the engine was a mount for the bottom of the triangulated rear swingarm.

Once rear suspension became universal, frames began to change dramatically. Adapting a suspended rear fork from a solid one seemed prone to failure.

Plunger-box springing allowed the rear wheel to move within its space in the rear fork, but the system was limited. It was the pivoted rear fork which began a revolution of sorts. The pivoted rear fork, essentially separated the two triangles of the diamond, giving the motorcycle two distinct halves. Although the separation was necessary, it created a nightmare for the overall design. Engineers have struggled with the inherent weakness ever since.

Various forms of damping, as well as different ways of mounting the engine, have lead to frequent breakthroughs. One of the earliest and best was the Norton "Featherbed" frame, which gave superb handling and good rider feel. The Featherbed was a duplex loop frame formed of Reynolds 531 chrome-moly tubing with a short triangulated rear section at the back of which were mounted the tops of two spring dampers. The bottoms of the dampers were connected to a pivoted rear fork.

As suspension performance and reliability was addressed, frame technology focused on the frame's structural integrity. The main goal of the frame became keeping the engine in the proper place in relation to the front and rear suspension. Rigidity was key. Designers moved away from loop frames to space frames.

Beam frames, and even monocoques with their thin-walled box structure, became fashionable again in the 1960s and until recently continued in the experimental stages. Clearly the top, or main front-to-rear support, of the motorcycle was its most critical design focus. From the Vincent twin to the Elf frameless bike, design-

Norton's racing guru, Joe Craig, stands beside the superb featherbed-framed Manx. Mick Walker collection

ers have played with this critical frame element. Some designers, like Kawasaki in 1980 with their Grand Prix bike, used the fuel tank as a means of stiffening the backbone of the machine. Many also have used huge single tubes, with various triangulated forms adding strength in critical areas.

In the mid-1980s, carbon fiber was utilized as was box tubing. Some of the designs even reverted back to double loop frames that were startlingly similar to the originals of the early-1920s. Currently, all teams at the 500cc level use the box-section twin spar layout. At least until something better comes along.

"When we started this [chassis building] three years ago, sometimes on the grid there were only twelve machines," Serge Rossett of ROC Chassis explained. *"When we started the last race we had thirty-six machines. We are very* *pleased with that. These bikes for customers are now competitive. We have seen for example, Niall MacKenzie is seventh in the championship with the private bike—with a private bike, not a factory."*

For state-of-the-art machines, Grand Prix motorcycles are not as high tech as their four-wheeled counterparts. In IndyCar, for example, the engine is stressed, actually becoming part of the chassis. As previously mentioned, that idea was tried by the Elf Team several years ago with little success. But it may have been the other components which ultimately doomed the bike, or perhaps the combination of too many untried things. Nevertheless, the frameless design was abandoned. Some think, however, that it will be back.

"There were absolutely no problems," said Rosset with 20-20 hindsight. And perhaps he is correct; perhaps the problems were with integrating several different revolutionary components, each requiring a great deal of scrutiny in its own right. But many feel that the next major breakthrough will be in this area of the motorcycle. "I think this is the way to go for the future. The way to go to a bicylinder, maybe less than 500cc to have the rpm and 100 kilo in place of the [current 130 kilo limit], because we see at the moment that at some circuits that are very tight corners the 250 is not quicker, but very close to the 500. So we have to reconsider for the future the size of the wheel. The main difference between the 500 and the 250 is the corner speed due to the weight, due to the size of the tires. At the moment we need large tires because we have the power. We can use the power, but you have to think differently for the future."

The 250s have been increasing lap times each season, while the 500s have been slowing down. The

This 250 chassis is much smaller and lighter than its 500 counterpart. The smaller frame would never accommodate a 500 engine, and the spars would twist like wet noodles under the power of a 500.

bigger machines have an awesome power advantage over the 250s, but the suspension and frame seem to be unable to assimilate the power, instead making the bike a rocket in a straight line and a bull in the corners.

"We've seen tremendous gains this year in the smaller classes lap [time] wise," said Erv Kanemoto at the end of the 1993 season. "We've seen increases of on average of a second from last year in the 250s and some of that is probably due to the competition. Others are from the fact that they seem to be able to ride the thing fairly hard. The engine power is also going up and they can run quicker in the straightaways as well. In the 500s, they are a handful first off. So there may be gains in some areas, but I'm not so sure they are able to utilize these gains in other areas of the race track. Or they are just more difficult because of the weight—as the weight has gone up it's become difficult to make the thing corner.

"We know that if you were able to build a light motorcycle—I think that something around 400cc or so would be an interesting project—I believe the corner speeds would go up," Kanemoto said. "But it's not a simple thing, because you would be able to use something like a 250 tire and you couldn't just use a 500 tire. So the tire companies would have to become involved. So it's not something you could just build and say, okay, we're gong to make this machine work. You'd have to have the tire people and various groups involved in the project. But right now, it seems to be at the limit in the 500s because they're accelerating so hard corner to corner. Then you have to stop the thing. It's harder to get through the corners. Although they're smoother to ride, I would assume the corner speeds may be affected. You're accelerating between the corners and when you get there you can't get through the corner quick—or any quicker I should say."

The reluctant consensus seems to be to make it handle by making it smaller and lighter. To do that it makes sense to decrease the engine size, which in turn makes the engine less powerful.

"And this is the way to do it," Rosset commented with a wide grin. "We have a possibility to build an engine especially made for that. A normal crankcase is

The Elf X was a mass of innovation—which may explain its failure. The engine acted as a stressed member; there was no frame per se.

not able to accept the stress from the frame. We don't make our own engine. So that means the Japanese making an engine built specially for that. To replace the frame is really quite difficult. That means we make things complicated to resolve the problems. That means we have to turn around one engine not specifically designed for this effect. But it's absolutely possible. We have done this in the past with the Elf Project. It was a bike deigned like the McLaren Formula One. So that means it was a single arm front and rear with a totally different concept. But to build also the engine. The casting must be different. You must have special material in the right place to fix all the accessories on the casting of the engine."

"If it continues with two-strokes (as opposed to a rumored proposal to eliminate two-strokes in favor of four-strokes), I think it would go that way (with a Cosworth type chassis). But I certainly believe they'll get smaller. And probably with a lot more exotic materials—like ceramic for instance," said Eddie Lawson. "If it goes that way you could do away with radiators, cooling lines, and water jackets The cylinders would get really small; the engines would just get really tiny. So I could see where things could get a lot more compact, lighter, and smaller."

Erv Kanemoto, Lawson's old tuner, was a little more cryptic, but suggested he agreed with Lawson. "I can only speak in generalities because I'm aware of what they've done, and it wouldn't be good for me to give you all the information. But I can give you the general things about it. One is that the expense is so great. Rigidity is a problem. The engineering staff can make the horsepower, but in the past they had to sacrifice some of the other things they believed in. Some of those things influence why the engine is the way it is or why the machine performs the way it does."

But as intelligent as the concept appears on paper, the standard way of approaching the problem has never been to reduce power. Rather the opposite is true. Moreover, the entire motorcycle must be reworked so that it has increased power—that is, as a smaller engine it must be at least comparable to the bigger 500—as well as increased rigidity. There are some design aspects of the motor that would have to be totally reinvented. It might be an expensive order to fill.

"At the moment we have a rule where we must have a minimum weight of 130 kilos for a 500cc. So it's useless to go in a way that's very sophisticated—like a bike with no chassis—using the crankcase as the main

The Elf E during test runs. Notice the ungainly front suspension. The bulk added to the bike's weight and created airflow problems as well.

part of the frame," said Rosset. "We have done this in the past, but it's not easy to work on the bike because we very often work on a two-stroke engine much more than a four-stroke. Each session we have to open the cylinder heads; we have to look at the cylinders because we don't have injection. So we need to adjust the carburetion compared to the color of the piston and other things and all these things. At the moment the engine can be easily put out of the bike and in the bike. We must not fix anything [on] the cylinder head or that part because we also change the gearbox very often. So at the moment, the most simple way is the way everyone else is using. To work easier in the engine. But if for example, we want to make a 500 bike that is two-cylinder, in this case we have the possibility to have a bike with a minimum weight of 100 kilos—thirty kilos less."

Rosset mentions another prohibitive factor: Honda bought all the patents and still has exclusive use of all designs. Designers wanting to experiment with what already exists—even Rosset—have to deal with Honda prior to launching a similar design. That may be a hard pill to swallow.

"I have no right. I don't work any more for Elf. I don't work for Honda so there's no way," he said in 1993. "We made forty patents. That means we made a full term of the problems and we locked the situation. It is difficult to make without having an agreement with Elf."

Some people, on the other hand, don't care.

"Serge has got a lot better idea of what everybody is thinking," Kevin Schwantz said thoughtfully. "I think the bikes are fine right now. We're not having any big problems riding them, and we're getting the chassis to take the stress and the strain of the power that the engine is putting out. The 250s are getting closer as far as lap time go but what that tells me is that a lot of the tracks are getting short. If we went back to a lot of the big, fast tracks like Philip Island or Hockenheim, the 250s are ten seconds slower than we are. If they'd just build fast tracks that were safe . . ."

But Schwantz is one of the best in the world. And eventually, the bikes will get to a point where even he won't be able to ride them. One of the toughest things for a fast rider to learn is that always going fast is not always the quickest way around the track. Sometimes sacrificing speed in one area makes gains elsewhere. Lighter and less powerful might, in fact, be better.

Above, *stripped, the Elf E appears as the simple design it was created to be. However, the sheer amount of adjustments which had to be undertaken made the project a constant game of catch-up. Eventually it was scrapped. Nevertheless, it was a fascinating project that motorcycle designers watched closely.*

Right, *Honda patriarch Soshiro Honda with André de Cortanze, the Elf Design engineer. de Cortanze hailed from the automobile world, where he had headed development for Ferrari, Alpine and Renault, and Peugeot. His background in automobile chassis design led him to radically revise the concepts of the motorcycle's front and rear suspension assemblies. The key goals were to preserve the attitude of the machine under both acceleration and braking and to lower the center of gravity.*

Honda's company holds dozens of patents which were developed during the Elf's tenure. Honda poured a great deal of money into the Elf project and still holds patents on many of the engineering innovations. Ultimately, some of the creation's components made it to the street, but largely vanished from the track.

CHAPTER 5 TIRES

Any given Saturday night at any one of myriad state fair or local flat-track circuits around the United States, an impromptu training of sorts is being conducted.

On those dirt tracks, where the leaders may swap the front spots more than fifty times in a twenty-lap race, road racers learn priceless lessons about moving a motorcycle sideways as well as straight ahead. Racers on the dirt tracks learn to slide a bike and at the same time keep it moving down the track without losing momentum.

More than any other component, the tires have been responsible for the evolution of the contemporary race bike. As the tires are improved, the frame, suspension, and engine struggle to keep up.

If the other three non-tire areas are developed more quickly, the improvements will prove counterproductive, because the bike can't outperform the tires on which it rides.

Dunlop may have introduced the radial tire to the circuit, but it was Michelin that made it work. Traditional bias ply tires created a great deal of internal heat as the rolling resistance heated the tire's woven cord. The challenge was in trying to create a tire that had good flex in the sidewall to act as a damper—a trait of bias ply tires—but with a stiff tread area capable of maintaining its shape under hard cornering. The result, eventually, was the radial tire.

Each team has a tire company race director who is responsible for that team as well as, perhaps, one or two other teams. Thursday afternoon he sits down with the team engineer to discuss the available tires. Here a gopher returns to the garage with newly mounted rubber.

The lessons paid huge dividends, because starting in 1978 with Kenny Roberts' first World Championship, up to 1993 with Kevin Schwantz' title, an American has captured the top honors every year but two (and in one of those two years where a Yank didn't win, a dirt-riding Australian, Wayne Gardner, did).

This loose and fast riding style isn't a secret. Racing of any kind is a learned skill. If a rider puts himself in a situation enough times, the reaction to a particular problem becomes second nature and he learns to deal with it subconsciously.

Dirt trackers have experience with the loose-surfaced tracks and are constantly searching for grip. They ride by the seat of their pants and have a good feel for the grip the bike has and exactly how much the power the track will allow.

Kevin Schwantz' all-wheel drive Porsche Carrera-4 split its torque quickly between all four wheels, delivering up to forty percent to any one wheel at any time. At any given moment, Schwantz could tell you where the power was being directed. It was one of the skills that helped catapult him to the World Championship.

Although the equipment in Grand Prix motorcycle racing is as sophisticated as in car racing, the techniques and the overall packages lag behind automobiles several years. We just discussed Porsche's C4, the all-wheel drive version of the 911. But the older rear-wheel drive 911, which is now nearly obsolete as a race car, was once in a league of it's own due to a certain design characteristic now considered a design flaw. The rear-engined 911 had a propensity to oversteer horribly. Race car drivers used that oversteer to rotate the car and essentially steer it quicker than the driver could with proper steering input. The driver would barrel into a corner, step off the throttle allow the back end to rotate and then bury his foot back in it. As the car began to spin, the driver would catch it with the throttle and in a blink of an eye the car would be headed the right direction and on its way out of the corner.

Essentially, the same conditions exist in motorcycle racing. But where snapping off the throttle in a 911 allows the back end to come around and point the car in the right direction, a violent shift of weight like that would devastate a two-wheeled vehicle. So road race riders use the application of the throttle to shift the back wheel. Again, like dirt trackers, the power provides the slide.

Dirt trackers blast down the straight, keeping an even throttle at the entry of the corner, then at the last moment flick the bike sideways and, with power still applied, allow the back end to drift out. If the bike isn't sliding correctly, it ends up in the loose stuff on the outside of the track because the line is essentially very wide with the apex. But since the bike's rear end is constantly stepping out it, it oversteers it into the corner and keeps it going the right direction—and at a nice clip, too.

So if the bike is going to be chucked sideways with controlled spinning and sliding of both the front and rear end of the bike, it's obviously going to come at the expense of some component of the bike. That component is, of course, the tires.

The one thing most people who occasionally watch the Grand National dirt track series generally don't know is that down on the track in any given corner, is a nice layer of rubber. Like the black streak at the outside of any paved corner, the rubber from tires is wiped away as the wheel slips and slides, leaving a nice layer of sticky black crud in the corner. The tire is essentially corruptible and is in a constant state of deterioration. Take that same movement from the dirt to the pavement, and the sideways slide of the bike is not nearly as pronounced as on the dirt. Yet it is every bit as important.

More than anything else, the tires have been responsible for the evolution of the contemporary motorcycle. The tires are the weak link as well as the inherent strength of the bike. As the tires improve, the frame, suspension, and engine struggle to keep pace. If the other three areas are developed more quickly, the improvements will be counterproductive, since the bike can outperform the tires on which it rides.

The history of tires really started with the man who changed the way the tires were used: Kenny Roberts. When Roberts applied his flat-track riding style to the pavement, the tires bore the brunt of the punishment. Eventually, his tires changed as his one-man relationship with Goodyear Tire flourished.

To back-up a minute, tires for bikes had always had grooved tread. The fronts were generally a ribbed pattern designed for stability and cornering, with the majority of the lines going the same direction as the tires—that is, circling the rim. The backs were designed for traction and featured a diamond-patterned tread. By today's standards, the rear tire was just short of useless.

But in 1973, shortly after Goodyear Tire had climbed to the pinnacle of IndyCar racing, the Akron company began to experiment with what seemed completely obvious to them as the best way to design tires: as slicks. They had already had a great deal of success five years earlier in IndyCars when they had completely eliminated tread. If the idea was to increase the contact patch of the tire, it made sense to get rid of the grooves. Rain tires were for rain—and had grooves. Dry tires would be slicks. So said Goodyear.

Most riders disagreed, so Goodyear tried another tack. Rather than go with full slicks they convinced Gene Romero to use a pair of fronts instead of a front and a rear at Daytona. Romero was the last rider to qualify on the 2-1/2-mile oval, and he trounced the field, easily

Kenny Roberts designed an extremely successful hand-grooved rain tire. So successful, in fact, that when Goodyear pulled out of the World Championship they took the design with them, eventually using it for Formula One. The tread design became dominant in auto racing and eventually filtered down to street cars becoming the successful, directional "Gatorback" radial.

qualifying for the pole. Goodyear beat Dunlop and riders began to pay attention to the Ohio-based company.

But the look of the tire didn't really change all that much. Or rather, it went backwards a little. The design went from being a round, balloon-looking thing to a nearly triangular slick with a single groove—actually more like a slit that ran around the circumference of the tire—which was supposed to allow maximum traction in a corner. Nevertheless, riders liked the fact that Goodyear had been innovative. They also liked the expedience with which they worked, and their ability to get new product to a new rider almost immediately. It captured the imagination of Roberts and eventually a famous partnership was born.

"We made tires for Kenny Roberts. That was one of my best projects," remembers Leo Mehl, president of Goodyear Racing Worldwide. "When Kenny went to Europe, we went over there against Dunlop and Michelin and we kicked their ass. They didn't expect it from Kenny. They didn't know what they were going

"The rider wants feeling," Dunlop's spokesperson explained. "With a tire that gives him feedback so that he feels it, and it doesn't sort of let go. At that stage he's off the bike. It's important that he has some soft feeling in the bike. Maybe it slides, but it slides progressively, and it lets him know what he's doing. It's a thing we've always been good at."

Tires are available in 190/55s on 6in or 6-1/4in rims. Some riders prefer narrower tires depending on the bike, the track, and so forth. On the front, most teams use a 125/55 on a 3in or a 3-1/4in rim. Rain tires range from slicks with a few cuts in them to full-wet tires. All are hand grooved, with the exception of the intricately designed full-wet tires.

to get. We had never raced in the rain because they don't do it here—they didn't do it then. We went over there, and we just took our dry design and our rain compound—we've always been really good at making rain compounds. We had a rain compound for sports cars that just dominated, and we put it on this rain tire and went over there and just kicked their ass." Mehl clearly relished the memory.

"I got the contract from Goodyear for Kenny," Mehl continued. "What finally happened was that Kenny had a little van he ran out of Holland or someplace, and we put a little trailer on the back and he'd carry the tires back there. I'd send an engineer to each race, and he'd [Roberts] tell him make the tires like this or like that or whatever and we made them just for him."

Which, as the Goodyear engineers were to discover, was not the easiest task. Roberts ran the 1978 season with Goodyear tires—he was the only Grand Prix rider on Goodyears. So not only was the compa-

ny making tires for one team, they were making tires for one man whose style was about as unorthodox as had ever been seen.

"You know how he leaned the thing over, that's how we developed them. They were made specifically for Kenny," recalled Mehl. "He leaned down farther than anybody. When you watched anybody ride a bike, nobody leaned over like him. They all do it now, but in those days nobody did. It drove us crazy, because we designed the thing so you could use the bottom of the tire. He used the whole sidewall. He was burning off the sidewalls. It was really him who designed the tires. You just look at him and you say, That guy's crazy. You just asked, Why didn't the bike go over? He was the first guy to wear a knee pad, to wear his shoes out; he was so much better than everybody at that time because of his technique. It's not like you have a car around you with a roll bar."

Roberts went out in 1978 on two different size

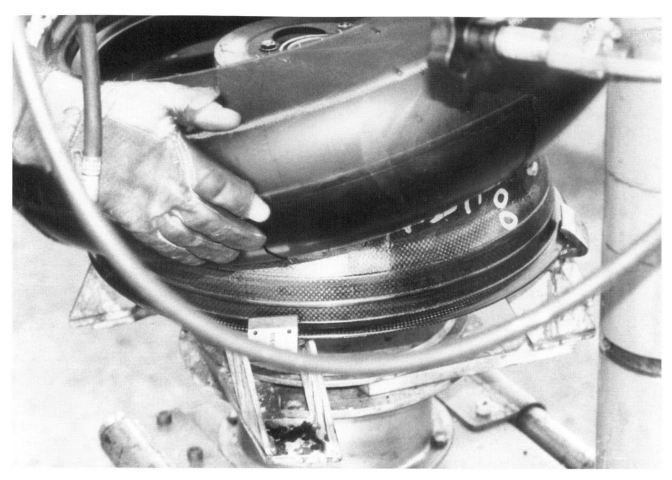

Cagiva, assisted by Ferrari, equips its bikes with carbon-fiber wheels. The ultralight components are not widely fitted due to high cost. They are treated as normal alloy wheels. Here a Michelin technician fits a new rear tire to a Cagiva rim.

bikes, contesting both the 250 and the 500 class. He won his first race on a 250, came in second on the 500 at the next, and until mid season won at least one race every outing with one of the two bikes. The performance was all the more spectacular considering Roberts had never raced at most of the European tracks. Roberts eventually dropped out of the 250 class, preferring to concentrate his efforts on the 500 championship. He won the World Championship that year and then dominated the next season with five wins in twelve races. The final championship with Goodyear came in 1980 when he squeaked by Randy Mamola for the championship after having a difficult second half of the season.

Goodyear got a great deal of publicity out of the partnership, but it was to be dissolved at the end of that season. "That was the best deal I ever had . . . with Kenny; he won three championship. I would say the first two years it helped him, and the third year we hurt him. Eventually Goodyear couldn't do it because we didn't make motorcycle tires. Kenny and I agreed mutually that if we couldn't step it up and really keep up,

that we should just quietly go away. But for a while, we had them, really," recalled Mehl fondly.

The partnership produced an interesting byproduct which only serves as a footnote here: Roberts designed a hand-grooved rain tire that was extremely successful in bike racing. So much so that when Goodyear pulled out of the World Championship they used the tread design for Formula One, winning their first race hands-down at Zolder. The tread design became dominant in auto racing and eventually filtered down to street cars becoming the successful directional "Gatorback" radial.

The perception of Roberts' three seasons had been that Roberts was new and that his advantage came specifically from the tires. That was not the case, but it spurred Dunlop, Michelin, and Pirelli to intensify the process of road racing tire development. Where Roberts used to show up with the little trailer, Michelin and Dunlop began showing up with transporters and throngs of engineers. Goodyear was outpaced, and the new era of tire wars in Europe had arrived.

Tire warmers provide a mass of spaghetti looking extension cord in any given Grand Prix tent and at any given race. Generators constantly hum, signifying the incessant need to warm the tires. Tire warmers—or tire blankets—keep the tires near operating temperature, permitting riders to get up to speed more quickly than on cold tires. Usually, the blankets get the tires to within 20 degrees of optimum temperature; the remaining is accomplished within a lap.

Tire manufacturers usually supply six to eight front tires and eight to ten different choices in rears—that's in slicks alone. The variations will be in sizes, profiles, compounds, and constructions. There is no "qualifying" tire.

Tire Options

Tires now are made in a couple of different formats, basically differing in the dimensions and profiles. The rears are 190/55s on 6 to 6-1/4in rims. Some people prefer narrower tires depending on the bike, the track, and other conditions. On the front, most teams use a 125 on a 3 or 3-1/4in rim. Rain tires range from slicks with a few cuts in them to full wet tires, with intermediates in between dependent upon the amount of water on the track surface. All are hand grooved, with the exception of the intricately designed full-wet tires. If the team is doing development for the manufacturer in its home country, the tires will be free—in a worst case scenario. Sometimes a fee is paid to the team for the advertising value of a winning GP team. Privateers also come to an agreement with the home manufacturer, but money usually changes hands—from the privateers' hands to the manufacturers' hands. Obviously, it works better when the team has a relationship with the manufacturer and can have tires specifically built for their riders.

"The rider wants feeling—a tire that gives him feedback so that he feels it, and it doesn't sort of let go," Dunlop Tire's Jeremy Ferguson explains. "At that stage he's off the bike. It's important that he has some soft feeling in the bike. Maybe it slides, but it slides progressively, and it lets him know what he's doing. It's a thing we've always been good at. What, perhaps, we haven't been so good at was the actual grip level. But now we've got the grip level up to a pretty competitive level as well."

Tires wars—or more specifically, tire choices have become the major focus of Grand Prix motorcycle racing. Tires from even a few years ago are obsolete.

"A lot of the material—the compounds, the ply materials, shapes—everything has changed," says Ferguson. "The rate of development is such that three years is a very long time. There are, historically, very big landmarks. Like when we first introduced radials. But those landmarks come every twenty years. Its more evolution not revolution."

Dunlop may have introduced the radial tire to the circuit, but it was Michelin who made it work. Introduced in the early 1980s, radials have become the way to deal with the breakdown of the rubber compound as a result of heat. Traditional bias ply tires created a great deal of internal heat as the rolling resistance heated the woven cord of the tire. The way the cord—which was once made of cotton, but ultimately formed with rayon, Kevlar, or some form of polyester—was laid down in the mold was important to the way the tire performed. The angles at which the cords crossed each other determined tire stability.

Rims vary in design and size as well. Compare this with the Ferrari-designed carbon-fiber wheel seen earlier.

A motorcycle tire has to deal with weight transfer as well as the angle of the bike in a corner. So rather than simply dealing with a relatively flat surface, as on a car tire, the bike tire will be abused across the tire's entire width. From one side of the bead to the other, the whole carcass of the tire is utilized.

When the angle varied a great deal in relation to the circumference of the tire, the result was a flexible tire. If the angles were very similar to one another, the tire, or actually the casing itself, was stiffer. So the challenge was to create a tire that had good flex in the sidewall to act as a damper—which came from the bias ply tires—but with a stiff tread area to maintain its shape under hard cornering. The result, eventually, was a radial tire.

The old, triangulated tire left the sidewall flexible with a very nice contact patch on either side of the tire, but nothing in the center. The contact patch in a straight line was nearly nonexistent, and the tire's tracking ability was relatively poor.

About the same time Goodyear was beginning to see wisdom in a partnership with Kenny Roberts, Michelin was beginning to see the wisdom in making a radial for Grand Prix racing. Having been credited with the radial revolution in street car tires in the late 1970s, Michelin produced the first motorcycle road racing radial in 1984 (Pirelli had actually marketed the first street bike radial three years earlier). The radial produced the desired results: a wide profile with a large, rounded contact patch, offering a flexible sidewall for a degree of damping. In fact, a good portion of the sidewall was eliminated while maintaining the same flexibility. The end result was a tire that weighed less, distributed latent heat better, and had much more predictable wear. At the same time, it was more consistent. More than anything, it provided a consistent slide, which the riders had sought since 1978. Tire development rocketed forward after 1984.

To recap: The structure of a modern tire allows the motorcycle to roll, it acts as a crude damper of sorts to take up the shock of small bumps, it keeps the machine tracking straight, and it keeps it stuck to the ground. At the same time, this tire is expected to steer the bike at high speed as well as gripping in low speed corners. But unlike an automobile tire, the footprint moves about the tire as the machine moves around the circuit.

The Tire at Work

A Formula One racing tire designer can count on tires being on the ground at all times. If the tires leave the ground, something more important than new tires are likely needed to get the car rolling again. Sure, the pressure on each wheel and the punishment any corner of the car gets varies at any given turn on the track of that particular week, but for the most part it is a two-dimensional set of actions. A motorcycle tire has to deal with weight transfer as well as the angle of the bike in a corner. So rather than simply dealing with the relatively flat surface that is a tire contact patch, where the footprint will vary only a few percent one way or the other, the bike tire is expected to be abused across the entire tire. In other words, from one side of the bead to the other, the whole carcass of the tire is utilized.

Consider for a second how the tire works: At a 90 degree or greater angle corner—say at the final turn at Laguna Seca (which is actually about 105 degrees)—the bike is under heavy braking. The contact patch, which is normally about the size of an average man's palm, has now diminished to almost nothing at the rear and contact is almost entirely on the front. The rear tire may be jumping across the surface of the asphalt, or chattering, as the back wheel is braking with little or no weight on it. The rider will lean back, ease off the brakes, ease on the throttle, and lean the bike

"A lot of long radius corners [are good for sliding the bike]—where it seems like you're in the corner forever on the side. You'll want to gas it to square the corner off," said Randy Mamola. "Once its starts to do the turn, you gas it.

Some corners are slow enough, and you're slow enough, to where you're not going to want to spin it." Perhaps. But, having watched Kevin Schwantz and Mamola over the years, not likely.

into the corner. As power is applied, the weight will come off the front, perhaps lifting the front wheel off the pavement, and the rear end will take all the weight as well as all the power. The rider will apply the power, perhaps slide the rear wheel, put the front back down, and off he goes toward the next corner. At least that's the simple description. What actually happens is much more complicated.

Randy Mamola was out of a job in 1993—at least out of a job as a competitor. Mamola, with his ability to not just spin a wheel—but to actually smoke a tire while exiting a corner—took a job with his old nemesis Kenny Roberts testing Team Roberts' Dunlop tires. Who better to test than Mamola?

As it currently stands with the FIM World Championship, once the season starts no rider can test on any of the sixteen tracks of the championship series unless a race has already been contended there for the season. Each team has a designated track where they can test at any time. Roberts' "home" track happened to be Barcelona. With Mamola, Roberts can test his bikes and tires on a track that has yet to be raced without either putting the riders at risk or jeopardizing the title.

So Mamola started in what was supposed to have been a week's worth of testing. Discovering how good he really was at giving Dunlop engineers feedback, Mamola has found a new calling—that of tire tester.

"I went to do some tire testing with Dunlop at a

track where they use sprinklers to wet the track down to test rain tires in Huntsville, Alabama. They have a factory there that produces 30 to 40 thousand tires a day. They have a testing ground with sprinklers on it so they can wet the ground and test car tires or motorcycle tires. And that turned into a full time position testing tires for them."

Mamola's normal work day would probably include a test of perhaps twelve to fourteen front tires. Maybe three will prove to be good designs. Then out of those three, one of the Roberts riders might try one the following week wherever there's a race. If there was time, Mamola might encourage Dunlop to build a new tire based on one of those three.

"What I'm concentrating on is what the tire's doing. So I go and weed through all these fronts, and that's been really interesting to me. I don't know how the tires are made, what they use and so forth, but it's been really interesting to have them put a new tire on my bike—say it's something totally out of the ordinary; something where they're trying different materials or something—and say, 'Be careful with this, we're not sure what its going to do.' I'll go out there and it's really interesting to come back in and tell them something and then they'll say, 'Well, that makes sense because . . .' and then they tell me what they've done to it. And its really interesting to try to understand. There's not a lot of extreme differences, believe it or not. I mean, there's some tires that you say, Don't ever put it back on. Forget it. It's wrong.

"What they do, is out of the fourteen tires there might be eight different constructions and they will all be the same compound. So you're dealing with the same compound, so all we're doing is searching for feel with the construction—in the different strengths in the sidewall, the way that they've put the tire on. If you can visualize this being the rim," Mamola holds his forearm up parallel to the ground and points to the top of his arm and he says this, "and this being the tire," pointing to his elbow and the underside of his forearm, "like if you're turning the bike, the tires collapse a lot.

"If you've ever seen racing in slow motion the tires collapse a lot. Its actually part of the suspension. We're braking straight up and down, and there's so much force being put into the tire and there's so much traction that they collapse."

In fact, the tires collapse so much that black marks from the sidewalls will appear on the rims. That means the tire is trying as to roll right off the rim. And to a certain extent it does. Fortunately, the combination of bike and rider is light enough that it doesn't actually accomplish this. It stays on the rim, and as the weight transfers off the wheel which is flexing and onto the other wheel, the tire gets a well-deserved rest. Until the next corner.

At some circuits, like Hockenheim, Germany, a flexible tire is useful. A flexible tire would be handy there because the bike's suspension would be set stiff in order to sustain the ultra-high speeds at that fast circuit. The inevitable bumps will be absorbed by the tires. As they rotate at 200mph, the centrifugal forces help bring the sidewall to its peak, thereby acting as a sidewall bolster.

For most tracks, however, the focus is not as clear.

"Sometimes if you have a very heavy-steering bike, you might use a harder, stiffer [tire] construction to keep that out. If it's a rough track, you might end up using a stiffer construction," Mamola explained. "But when you're racing, you don't look at a circuit like that, you just go out and say, 'I'm having a problem with the front end chattering here or there,' so they might change to a different construction. They might say, 'Okay, we're going to put this one on,' and I just go out there and try it and I say if its better or if its not. Whereas when I'm testing, I try to find the area where I have a certain problem with the tire and I try to decide, 'Yeah, this is going to work at this sort of a race track, and this tire's going to work at that sort of a race track.'

"A lot of tires we've used this year (1993) have been one particular front. There's so many things that come into it. The tire temperature, the track temperature. I mean, you're talking 55 degrees centigrade which is a great amount. We're talking 120 degree tire temp. Each affects the way the tire wears, each effects your decision on what tire to run. But for the conditions, they already know, basically, because of the past zillion years of racing, what works and what doesn't work."

Adhesion has gone up each year, and because of that engineers need to strengthen something else. As we discussed earlier, the tires are usually the catalyst of change. The better the tire works, the more developed the chassis has to be, the more efficient the suspension has to be, and then the brakes have to be improved, and so on.

But so far, the history of motorcycle racing tires has been preoccupied with the way the grip diminishes. As in all forms of racing, the ultimate tire would be the one that has incredible grip and which lasts forever. In motorcycle racing those goals still exist, but are

No rider can test on any of the sixteen tracks of the championship unless a race has already been contended there for the season. With Randy Mamola (right, with hat),

Roberts can test his bikes and tires on tracks that have yet to be raced without either putting the riders at risk or jeopardizing the title.

not as important as feel. Tires deteriorate—it is a fact of racing. How consistently the tire deteriorates is a major concern. Riders want to know how far they can push and for how long. But more importantly, they want to know when the tire will go from stuck to un-stuck—at whatever stage of deterioration the tires are in. Not only is the feel important for safety, its also crucial, as we discussed earlier, to performance.

"There's some tires that you stick on and, say the race is only thirty laps, you might be able to go really fast in qualifying and for five laps (in the race) before it really starts to slide around," says Mamola. "Yet take that tire into the race for thirty laps, and you might be able to do the last lap as your fastest lap, if you were pushed to that point, because it's really controllable. That's what we're looking for, to be able to slide them around. If a 500 can't go around the corner unless you're on the gas after you've settled the front in the corner, [then] you have to get on the gas to get the thing to steer from the rear.

The tires collapse so much that black marks from the sidewalls will appear on the rims. The crew cleans them up prior to the next use.

"[How it steers] depends on the track surface. There are surfaces that when you look at them they just look like a polished-type pavement, and you know there's not going to be much traction. If you can see the video of Wayne [Rainey] winning the race in Barcelona (in 1993), you'd be able to see what I'm talking about as far as us sliding the bikes around."

In Barcelona, in 1993, Wayne pulled away quickly and with little opposition. The combination of fast and slow corners there make it a track where the riders need the ability to slide the bike when speed alone will not help. It was no contest between Rainey and Doohan. Doohan, who undoubtedly had the more powerful bike at Barcelona, could not keep up with Rainey. The reason was simple: Doohan didn't have the ability to finesse the bike around the curves by sliding his Honda. Where Rainey could spin the rear Dunlop of his Yamaha at will, Doohan had to push the Honda to get the wheel to slide. Usually he was past the limits of adhesion when the wheel unhooked. The result was a bike that tried several times to buck him off. Finally, Doohan had to settle in behind Rainey.

"The Michelin riders are having a hard time," Mamola commented with a grin at the end of the 1993 season. "We call it the 'Michelin flick.' Usually, it gets to a point where it's spinning and then it lets go

Kenny Roberts applies the power at Laguna's final corner and straightens up as the bike spins away. The front wheel is cocked ever so slightly to the right, indicating that the rear end is sliding a bit.

and the rider falls off. Dunlops have always been an easier tire to power slide and much more predictable. And so, therefore, we believe Michelin is trying to build a tire more similar to Dunlop's. I believe the difference is in the he material as far as construction and so on. Our tires don't jerk around normally. Ours are a lot more controllable and more consistent in a spin. Once you spin it, like I said, the Michelins, at some of the racetracks, will want to break away if you try to force it harder and harder. That's what Wayne was trying to do, to push the pace up to where [Doohan] would try to push that hard and he just couldn't go anymore otherwise he'd fall. "

Technically, what happens where the tire meets the ground is fairly complex. Rubber, like anything else, is made of a chain of molecules. The chain governs its adhesion to the particles of the pavement. It is also the reason that a slick, dry tire is virtually useless in wet conditions: those particular molecules are not designed to bond with wet pavement.

The compounds change at nearly every race. There is no such thing as A, B, or C tire choices as in Grand Prix racing. At each race, each team has a race director who is responsible for that team as well as perhaps another one or two teams. Thursday afternoon he sits down with the team engineer and tells him how many of what tires they have. Thursday they have a baseline.

"This week, for the Marlboro Roberts 500s alone we've probably produced more than 100 different types of tires," explained Furguson. "They're not all completely new, its, a progression. We take what happened a couple of races ago and the test we last did at the circuit and we say, Okay, this, this, this, and this need to improve and this is a variation of that, and this compound has been changed so its the same family of compound . . . that sort of thing."

At any particular race, the tire manufacturers will probably supply six to eight front tires and eight to ten different choices in rears—that's in slicks alone. The variations will be in sizes, profiles, compounds, and constructions. There is no "qualifying" tire. Although there is a push for the front row, the teams don't become preoccupied with the top spot because it is far less important than in auto racing where there are only two drivers with a shot at the lead in the first corner. In Grand Prix motorcycle racing, it is more critical to get setup for the race.

Tires from the beginning of the season are no longer competitive at the end of the season. The older stuff goes to the top national riders. And as the quality gets better in the top ranks, it gets better in the lower levels as well—although it is never as good or as competitive as it is with the elite series or teams.

Hot Treads

Heat—the ability to withstand it and the ability to generate it quickly—is constantly being experimented with. Heat changes the way the rubber grips the road, enhancing it to a point, but when overheated, the tires begin to produce diminishing returns. Oddly, the tire can lose adhesion on only one side or on one specific portion of the tire. The layout of the track is important in considering when and how the tires will begin wearing.

"When you first get on this track (Laguna Seca), you really want to make sure the right hand side of the tires are warm because almost everything here is left," says Mamola. "If you visualize it, everything goes left. A lot of people don't realize it and a lot of people fall here on the right side. You have two sides of the tire."

Something that rarely occurs to automobile racers is the relationship between front and rear wheels. The front and rear track can be offset—that is, the back can be wider than the front. The car is still essentially a rectangle. The wheels may change the balance slightly, but they do not change the geometry of the car enough to cause a great deal of trouble. However on a bike, the differences make a motorcycle more or less agile depending on the relationship between the front and rear wheels.

Think about that a minute. With a front tire that is 3 inches wide and a rear tire that is 6-1/4 inches, the wheel alignment goes from dead straight when the machine is headed straight, to offset as it leans over for the corner. So not only does the wheelbase change and the rake and trail alter under heavy braking, but the tire contact patch alignment is changing also.

"Just think about leaning the bike over," prompts Mamola. "You've got one wheel way over here—three inches wider. That's a lot." The way to make a sluggish bike more responsive is to slide it. And the key to sliding it is to get the tire temperature up."

To try to get the tires to work as quickly and as efficiently as possible, Grand Prix teams utilize tire warmers like those used in Formula One car racing. The electric warmers bring the tires up to operating temperature so riders can be competitive within a turn or so. Nevertheless, some riders still have their own litmus tests.

"We use our legs and our knees to judge how far we're leaning over," explains Mamola. "When we do, we basically know where we're at with the tire. In the rain I try to touch the ground as much as possible. Because if the tire ever slides, sometimes you can push to save it. Not a lot but sometimes."

Once the heat has built up in the tires, the rider tests to find its limit of adhesion. That will be essential as the race unfolds.

"You start just like if you were driving a car on gravel or snow. You'd accelerate in second gear, and you'd feel it start to spin you wouldn't give it more gas, you'd give it less—if you weren't racing. The same on the bike. You start to gas it, when it starts to spin, you hold. If it's got good traction, you just keep increasing it until it starts to spin and then you've already got it coming out of the corner. Then next lap you already know what that feels like in the particular corner. Turn four (at Laguna Seca) is one of those particular corners where you've got to get it to spin to get it to come out of the corner.

"To get on the gas on its side is where we have the least amount of traction, the least amount of contact area. If it's way on the side and we're coming back to this area," he says, again holding up his arm and pointing to the center of the underside of the freckled forearm. "If we're down here on the edge," he says pointing to the side of the arm, "because it's leaned over so far we can get the ground to be up on the [fiberglass fairings]. We've got more traction here. So basically, you want to gas it to get it to start to turn, and then the bike will start to pick up. You want to get it picked up as much as you can as early as possible because you can give it more gas when you are up straight. The quicker you can get in and out of the corner to do that, the faster you will be able to go around the track. The (tire marks) will start thin and they will get wide as we begin to pick the bike up. Because you're going from the sidewall, which has hardly any contact area to the flatter side of the profile.

"A lot of long radius corners [are good for sliding the bike], where it seems like you're in the corner forever on the side. You'll want to gas it to square the corner off. Once its starts to do the turn, you gas it. And what we've learned on videos is that once the tires start to collapse, the profile's actually changing to line up more to match the front. Some corners are slow enough and you're slow enough to where you're not going to want to spin it. Turn three is a good example. You barely spin it because you don't have a lot of room if you're going to spin off the track. Where you've got tons of room, you want to get on it early because you've got a straight. Therefore it starts to spin because it's on the powerband. It works out between the powerband, the speed of the corner, the angles of the corners—and of course, you're changing the angles of the corners if you're spinning it. All bikes work differently."

All bikes work differently, for sure. Not so obvious is that all riders ride differently. A great tire for one rider is a horrible one for another. And like any other component in the chain, the team must rely on the feedback of their man. If in every test and every analysis, the tire is better than the one it replaced, yet the rider feels less comfortable, then progress is halted. The idea, after all is to win. If the talent says its faster on a bad tire, the team shuts up and goes along with it—hopefully to the World Championship.

CHAPTER 6 BRAKES

Course marshals found the pin which had worked its way loose somewhere near Laguna Seca's Corkscrew. A half lap away, Eddie Lawson was being loaded into an ambulance and taken to the hospital. Lawson had ridden the Yamaha through five turns before the front brake pads finally popped out. Eventually, the brakes completely failed. He was forced to dump the bike in the dirt before crashing heavily into the hay bails. Lawson recuperated quickly, but by then the 1990 World Championship was lost.

A year earlier, Wayne Gardner had found Laguna Seca frustrating as well, to say the least, as he too had problems with brakes, having been pitched from his NSR twice in one weekend—both times from brake trouble. In Gardner's case it was more a matter of getting used to the equipment, since it was his first experience with carbon brakes. Gardner had found the stopping power of the new exotic-material discs—which were fitted for the first time on the Honda in 1989—almost violent. He announced shortly afterward that he would never use the carbon discs again as long as he lived . . . or at least until he had forgotten the California nightmare. Fortunately, racers have short memories when it comes to accidents.

But the incidents illustrate again how critical components are on a Grand Prix motorcycle. In a car, a brake failure means slower laps; driver anticipation of corners necessitates longer off-throttle time. But safety-wise auto racers can generally do without brakes, using the gears and the steering angles to scrub off speed and slow the car down. NASCAR drivers regularly lose the entire brake system on shorter circuits. They have to adjust their pit stops accordingly, but usually can remain in contention for top ten finishes. In Grand Prix bike racing, losing braking power will surely result in finishing well back in the pack. It can also result in fates similar to Gardner's and Lawson's. The circumstances were different, but the results were the same.

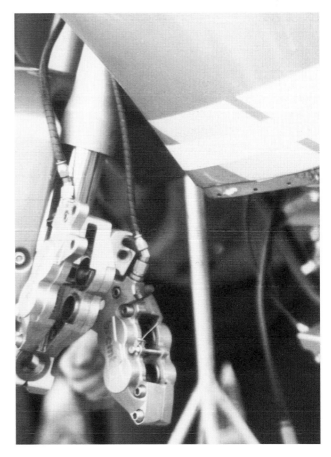

The front wheels are equipped with two rotors, each clamped by a four or six-piston caliper. Here the front calipers hang from their braided steel lines as the wheel has been removed.

Braking systems have made great strides in the past few seasons. Carbon brakes are now common, their performance now quantifiable. Their characteristics no longer surprise riders who once found themselves on their butts while coping with the learning curve. What happened to Gardner had to happen to

Before the advent of disc brakes in the 1970s, drum brakes hauled GP bikes to a halt. As speeds increased, so did the required stopping power. This led to massive twin-sided, *four leading-shoe brakes like these fitted to the Yamaha TDs and TRs. Mick Walker collection*

somebody. Gardner found out how quickly carbon brakes stop after generating some heat.

Carbon brakes had been created initially to reduce weight. The difference in weight between a standard cast-iron or steel disc and a carbon disc is almost fifty percent. That is, the carbon disc weighs about 2lb, where the cast disc weighs about 4lb. That in itself was sufficient reason for the teams to change from iron to carbon; weight savings are desirable anywhere, but taking pounds off an unsprung object is particularly critical. Also, the brakes rotate with the wheel, and the more fat that can be trimmed from a rotating mass the

better. Gyroscopic as well as flywheel effects both enter into the equation, and ridding the bike of excess weight seemed appropriate in 1989.

But the side-effects from the diet were not with the weight but with the material itself. Carbon brakes do not fade like cast brakes do. Not only that, carbon brakes actually work better as they heat up—a completely new concept for riders who anticipated with trepidation the point in the race where the brakes would be heated to a point where stopping was done on a wing and a prayer.

What finished Gardner was exactly the opposite of that—the increased ability of the carbon brakes as

The rotor material is often mistakenly referred to as carbon fiber. Actually, the rotors are constructed of carbon which is formed under heat and tremendous pressure to form a sandwich of material. The shrouds pictured here, which help keep heat in and debris and water out, are carbon fiber.

The carbon discs gain efficiency as they heat up. The effort to stop the bike after five minutes of riding is less than a third of what it was just a few minutes earlier.

Although this rear disc is carbon, cast-iron or steel are the more frequently used rear disc materials. Note how much smaller the rear disc is compared to the fronts.

The carbon disc weighs about 2lb compared to the cast-iron disc's 4lb. Reducing unsprung weight is always an advantage. Most designs have the carbon disc mounted to a carrier which is then mounted to the wheel.

they heated up. Gardner had gone into the first several corners of the lap at Laguna using the brakes as if they had been cast iron or steel. In the beginning, they reacted exactly as the cast-iron had. Perhaps Gardner had felt the difference in the way the bike handled, but other than that, on those first laps, the NSR felt almost identical in terms of stopping as it had on the irons.

But as the Carbon discs heated up they began to become much more efficient. The effort to stop the bike after five minutes of riding was less than one-third of what it had been those few minutes earlier. In fact, as the brakes were applied—even at a perfectly constant pressure—the stopping power on any single application would rise almost exponentially. In other words, if ten pounds of hand pressure produced a feel

that would stop the bike at, say, one G at 150mph, the same disc at optimum temperature would stop it with twice the power in half the distance. Of course, as the disc heated, the bike had already slowed. It's easier to slow the wheel from 80mph to 30mph than from 150mph to 75mph. But at the slowest point, the brake is at its best and it tends to clamp so effectively that the motorcycle can stand on its nose.

Saturday, Gardner had braked hard up to the final corner before Laguna Seca's main straight, then trailed his fronts slightly into the corner. As the brakes warmed, that trail—even just resting his finger in the lever—was clamping about the same as it had been under hard braking minutes earlier. The front wheel locked and Gardner went flying. The next day he went back to the steel discs.

The calipers are one-or two-piece and machined out of a solid aluminum or magnesium billet. Calipers are almost always mounted on the back side of the fork because it places that weight closer to the steering axis thus reducing inertia when the front end is turned.

The carbon discs work best at somewhere around 500 degrees Fahrenheit. The teams monitor disc wear and temperature at different venues and under different conditions by applying special heat sensitive paints that changed color to indicate temperature. The results are closely observed at tracks like Laguna Seca where riders often generate temperatures that make the brake fluid boil due to the extreme braking conditions.

Often called carbon *fiber* discs by riders, that description is not strictly accurate. The brakes are carbon. Carbon fiber is a composite—like fiberglass—which means bonding with resins and such. It is impossible to manufacture brake discs from carbon fiber since they heat so fast and to such a high temp that the bonding material—be it glue or filler material—would simply melt away.

Carbon discs are formed in layers and over long periods of time. Vacuum and/or heat is used to create the raw material. Originally it took five months to produce a Grand Prix motorcycle brake disc. It takes far less time now as the process has been perfected.

The unfinished discs are milled and ground to tight tolerances, destroying even the hardest tungsten tooling in the process. A warped carbon disc is as useless as an untrue cast one. As it is, the makeup of the carbon causes the disc to heat in some places quicker than other places. Not as common today as it was in 1989, the uneven warming creates an unsettling vibration. Usually it goes away as the entire disc goes through a few complete heat cycles. Otherwise it is tossed in the trash.

The carbon disc's size has changed as the years have gone by. Initially, they were made smaller than their standard cast brethren. The stopping power was the same as with the smaller disc, so why increase it? As discussed earlier, Gardner had a hard time coping

Opposite page, a set of six-piston calipers on a Suzuki RGV500. Although brake performance continues to improve, stopping power is limited by the overall design of the motorcycle and the ability of the rider to keep it upright during extreme braking conditions.

The front master cylinder's fluid reservoir is like a street bike's in function, though it is often larger to accommodate the greater fluid expansion caused by the heat generated by six-piston calipers clamping carbon rotors.

with the brakes increased power as it was. Increasing its diameter was not an alternative then. But as the riders grew accustomed to the carbons' characteristics the brakes improved and the discs grew in size.

"We've varied between 300 and 310mm carbon discs," explained Kevin Schwantz after his World Championship 1993 season. "At one stage we had them as big as 320mm and even went out to 330s. But at that stage we reached a point where I thought the brakes had so much initial bite that it was causing a problem. There was no transition from when you

Opposite page, metal alloy rotors on Wayne Gardner's NSR500. Initially, he liked the metal rotors far better than the carbons, feeling that the carbon brakes grabbed too severely. But like anything else, the carbon rotors just took some getting used to. Most riders now use carbon brakes.

were on the gas to on the brake. There was no time for the bike to settle. It wanted to immediately stand the bike up on its front wheel."

The front wheels are blessed with two discs. They are, depending on the circuit, varied between 290 and 320mm in diameter. In 1994, some of the less wealthy teams still clung to cast discs, but the major teams had all gone to carbon. A set of discs cost somewhere around $2,000 in 1994.

Like any modern motorcycle brake system, the rider operates the front brake with his right hand and the rear brake with his right foot. When the brake lever is compressed, it moves a cylinder which forces fluid through the brake lines. As with any street bike, the pressure forces a set of pistons in the caliper to move the brake pads into contact with the spinning disc. The resulting friction slows the spinning motion of the wheel. The master cylinder is adjustable by tightening or loosening a screw that changes the leverage ratio which will affect the performance of the master cylinder, which in turn increases or reduces the clamping power of the pistons.

The rear brake discs are generally still made of cast iron or steel and are very small compared to the fronts—about 210mm is the standard. The rears, because they are cast, are drilled for weight savings and heat dissipation. Drilling can be done laterally, axially, radially, or tangentially. The calipers are one- or two-piece and machined from solid aluminum or magnesium billet, set with two pistons with carbon pads. The pads and discs are of like material—although lately the pads have been imbued with a softer composition so as to wear quicker than the disc itself.

Nevertheless, the rear discs are almost redundant, since all the action happens at the front of the bike. The rear end is just along for the ride. For a lot of riders the rear brake is used just for feel. When the wheel is in solid contact the brake's effect can be felt; when it is hopping or nearly weightless, that condition is identifiable as well. The rider will know from that feel what the balance of the bike is and how much more front brake can be applied in any given corner.

The front end is equipped with two calipers usually machined from aluminum or magnesium billet—billeted pieces seem to retain the rigidity better than cast calipers. Many are cast with a metal matrix technique of quenching and cooling that imbues the piece with the strength of a billeted piece of metal without having to compromise the overall design to accommodate machining. Each caliper usually has four pistons for incredible clamping power when paired with the carbon discs. The calipers generally have different sized pairs of pistons, usually varying between 29 and 35mm in diameter.

As previously discussed, the carbon brakes must heat up to work properly. At the same time, the heat from the disc will boil the fluid if care is not used in allowing the disc to arrive at its optimum temperature. The calipers are almost always mounted at the rear or trailing side of the fork. This helps cool the discs, by placing the calipers out of their airflow. (According to Hon-

da HRC crew, by the way, there seems to be very little difference between placing the calipers either on the leading or trailing edge of the disc.)

Nicely formed shrouds were conceived to keep the brakes warm in the rain. At the same time, the shrouds tend to keep problems in a restricted area. If, for example, Eddie Lawson had had the problem he suffered in 1990 with a covered disc he probably wouldn't have been as lucky as to be able to dump it where he liked. What Lawson faced was a corner and no brakes. The pads had fallen out somewhere along the front straight. Had the shrouds been on, the pads more than likely would have wedged themselves into the works of the brakes with the end result having been a bike that suddenly stopped rolling and shot Lawson onto the pavement—perhaps at 160mph.

Brakes develop with the riders. Schwantz, a fit man, has complained about the punishment his body takes during the course of a weekend. Riding a Grand Prix motorcycle is difficult enough as it is, but as the engine makes the bike go faster and as the clamping power keeps moving up to keep up with the increased engine power, the human body will have to be recreated to cope with the forces inflicted upon it.

"It seems like it puts a bigger pounding on the body. I think our telemetry measured well over two Gs through the brakes many, many times during the season. That goes back through your arms and neck. Just trying to hold yourself on the bike is a pretty big pounding in itself as well as the mental pounding you take weekend in and weekend out."

Ultimately, brakes are what keeps the rider from flying off the road and into the concrete, hay bails, or whatever else lurks off the side of the racetrack. In the balance between riding on the edge and riding beyond it, brakes can become as difficult to tame as the power that makes the continued evolution of the brakes essential. "So the brakes, I think, are developing right along with the bike and the engine power," said Schwantz with a contemplative sigh, again recalling the punishment his body takes each weekend. "And I think they'll have to continue to do so."

At least as long as the engines continue to improve . . . and the riders continue to tame them.

As with most street bikes, a remote reservoir (not shown) feeds the Grand Prix bike's rear master cylinder.

CHAPTER 7 TRANSMISSION

In terms of transmitting power to the ground through the gearbox, Grand Prix motorcycles are slightly lagging behind what seems to be fairly significant four-wheeled technology. In the four-wheeled racing world, semi-automatic transmissions, paddle gear changes, and computer-controlled selection are becoming the norm. And although most Grand Prix motorcycle teams now utilize some form of momentary power suspension to facilitate gear shifts, for the most part, the technology is far behind that of automobile racing. Grand Prix bikes are still designed in the traditional layout: foot operated sequential shifter and chain drive. The system leaves room for improvement.

Although hand shift, foot clutch bikes described the original layout, the men who rode the pioneer machines insisted on a better design. Imagine being hunched over a heavy bike, manhandling it through corners, and then having to worry about taking one

Until the 1930s, handshift was de rigueur on racing bikes. The racer shown is a Moto Guzzi 250cc single from 1926. Mick Walker collection

Just as in a race car, the clutch is used as a buffer. In Grand Prix bike racing, the clutch also acts as a device that can be used to save a rider from the disastrous effects of a blown engine. Take a look at the rider's left hand; you'll notice he has one finger hooked over the clutch lever. Should the engine seize, the rider has fractions of a second from the time he senses the problem to the time that it seizes the rear wheel. Disengaging the clutch allows the rear end to free-wheel, hopefully preventing a crash.

In motorcycle racing, most short shifting is done because an extra shift of gears would waste time. Unsettling the bike doesn't seem to be an issue. "A lot of times you do short shift," explained Niall MacKenzie. "Sometimes it will be because we need to be in a taller gear for the next corner, but we don't necessarily need to accelerate hard up to that corner. So you just change from second to third early on so you're setup for the next corner. The other reason you short shift is because you'll come up to a relatively slow first or second gear corner and the bike wants to wheelie and the front wants to pick up. So you short shift into a taller gear so it won't have the power to pick the front wheel up."

hand off the handlebar to select another gear! From a lack of necessity and due to an inherent shortage of space, most early Grand Prix bike gearboxes had only four speeds. It wasn't until the Japanese invasion in the 1960s in the smaller displacement classes that the gearbox became a focal point for performance. Eight speeds were normal in the mid-1960s, fewer gears were considered uncompetitive. Since then, regulations have put a cap on what was becoming an expensive component. Six is now the maximum number of gear selections on a Grand Prix motorcycle.

The transmission of a modern Grand Prix bike is laid out essentially the same way as a street bike. The crankshaft is geared to the clutch which turns a mainshaft in the transmission. A shifting drum positions shift forks which move from side to side to engage and disengage the gears. Most bikes are fitted with what's called a cassette-type gearbox, meaning the shifting drum and shafts can be removed by sliding the unit out of the side of the case without having to separate the entire crankcase. It needs to be a hardy component.

Things are done with a car so as not to upset the its balance. Application of the brakes, of the throttle, shifting of the gears is all done so as not to create a change in the smooth, fluid movement of the machine in motion. But on a bike, there is little difference in overall performance if the bike is manhandled rather than finessed. Riders brake hard, accelerate hard, and shift gears as quickly as possible. The effects can be seen as the machines head down a long straight, bucking and skittering each time the rider shifts gears.

"I think, initially, the idea is not to unsettle the bike," said Niall MacKenzie. "But there comes a point at the exit of the corner where it doesn't really matter at all, safety-wise, and all that matters is getting the throttle open and getting the power down to the race track. So yeah, initially, it's important not to unsettle the bike from a safety point of view, but after that it doesn't really matter if the back end is stepping out a bit or not. You're really in a position where it's not going to throw you off. In fact, as long as the bike has contact, it's not losing you any time. I think that's

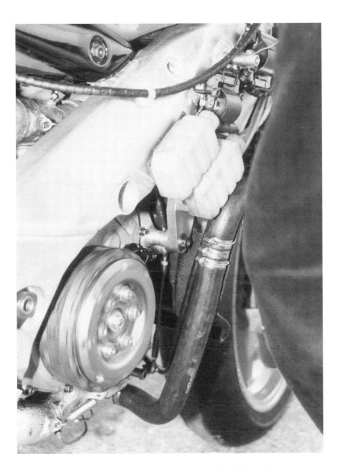

A dry clutch, like the one fitted to this ROC/Yamaha, is more efficient than a wet clutch, like that typically fitted to street bikes. Because the clutch is not immersed in oil, there is no drag due to oil shear (a horsepower loss with a wet clutch) and engagement is more precise.

"The chain will stretch a couple of inches—between an 1-1/2 and 2in," said MacKenzie. "That may be an extreme figure for one warm-up, but it's not less than 1in for sure. Once it gets to it's maximum, it won't stretch any more. Then you can use that chain for the race and on into the next practice."

Currently, all Grand Prix bikes in the 500 paddock use chain drive. Shaft drive has never made an impact in Grand Prix racing, and belt drive is currently far too inefficient.

probably what you see. It isn't particularly smooth sometimes the way some riders ride, but it is the quickest way to get through the corner."

In auto racing, drivers use short shifting both as a way to keep from having to upshift, then drop back down, but also as a way to keep the car settled. This is especially true when going through a fast corner where full throttle is not quite appropriate until the corner exit. But in motorcycle racing, most of the short shifting is done just because an extra shift of gears would waste time. Unsettling the bike never seems to be an question.

"Obviously, every racetrack has different corners and a lot of times you do short shift," explained MacKenzie. "Sometimes it will be because we need to be in a taller gear for the next corner, but we don't necessarily need to accelerate hard up to that corner. So you just change from second to third early on so you're setup for the next corner. The other reason you short shift is because you'll come up to a relatively slow first or second gear corner and the bike wants to wheelie. So you short shift into a taller gear so it won't have the power to pick the front wheel up. It means you aren't accelerating as fast, but you aren't losing as much time as you would if the bike was standing up on one wheel."

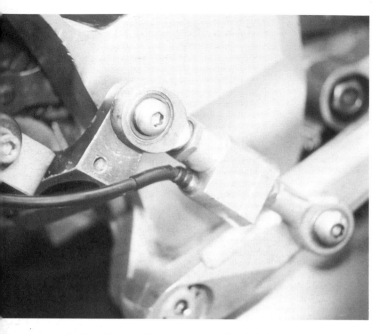

As the rider applies pressure to the shift lever with his foot, the sensor sends a signal momentarily killing the engine, allowing a quick, clutchless shift.

The clutch is a multi-plate unit. Here it is as it appears without bodywork covering it.

No concerns with the power unsettling the machine?

"You aren't feeling the vibration from shifting at all. It works particularly good now because most of the bikes are using semi-automatic shifters—as soon as you touch the gear shift it upshifts without touching the throttle; it just keeps the throttle wide open. If the front begins to come up, you shift and the throttle comes back down, whereas before when the front came off the ground, you needed to back off the throttle to shift. You had to come off the throttle which was fractions of a second longer. So it works quite well now."

So the biggest concern for riders was the loss of time in shifting, rather than the suspension movement which occurs as the bike makes the transition from power on to power off as shifting occurs. To fill the power gap, most teams have gone to a primitive semi-automatic transmission. In other words, the engine is killed momentarily, sort of the way Formula One cars were slowed during the days of traction control. The engine is sent a signal from the shift lever, telling it to interrupt power just briefly. The engine, making what sounds like a misfire, will allow the immediate selection of the next higher gear.

"When you touch the gearshift, it kills the ignition; it kills the carburetion. The only reason it shifts gears is that the engine cuts and it jumps into another gear. It's quick and consistent. It's crude, but it's not crude, because it does kill the ignition and the carburetion. It's not just a kill-button-type thing, it kills it for 0.04sec, I think. Just enough time for it to shift. You're

still pushing on the gearshift, so you're still mechanically shifting as it were," MacKenzie said.

Even with the electronic devices—that seem more like afterthoughts than actual transmission components—transmissions rarely fail during races. More likely, they simply begin operating abnormally. Selection gets more difficult, and movement in and out of gear becomes labored.

"Sometimes it's difficult to find neutral. Like, first and second get stiff, with a stiff mechanism," MacKenzie said. "Sometimes they miss a gear. Sometimes we get false neutrals."

The clutch is found at the end of the transmission shaft, at the point where the transmission and the crank are geared together. The clutch acts to engage and disengage the engine power from the gearbox. The common unit is a multiplate dry clutch with alternating driven and driving plates. The friction plates have a surface of sintered metal—essentially a metallic powder welded together. This type of material ensures a high coefficient of friction even when hot. When the rider engages the clutch, a cable is slackened which allows a lifting mechanism to move a spring-loaded pressure plate against the clutch plates. The typical clutch measure a mere 5in in diameter.

In Grand Prix bike racing, the clutch also acts as the device that saves riders from the disastrous effects

This multi-piece shift linkage allows for considerable adjustment to fit the rider.

of a blown engine. Take a look at a rider left hand; you'll notice he has one finger hooked over the clutch lever. Should the engine seize, a rider has fractions of a second from the time he senses the problem to the time that it actually seizes the rear wheel. Disengaging the clutch allows the rear wheel to free-wheel . . . hopefully preventing a crash.

The clutch in a race bike when compared to the clutch in a street bike doesn't seem to do much. Most riders use it only occasionally. Some use it only at the start, some use it to upshift, and some use it for downshifting only.

"I use the clutch for backshifting. It's just sort of an engine brake to me," MacKenzie explained. "I just backshift and feed the clutch out. You've got to be more careful in the lower gears when you're feeding

the clutch out, because it can lock the rear wheel.

"Some riders kind of blip the throttle so that they really haven't got a lot of engine braking. The engine is sort of catching up with the transmission. They obviously have to backshift. They [riders that blip the throttle] do that because they probably use more rear brake than perhaps I do. I tend to use the transmission more than the brake. I do use the rear brake, but there's more feel with the clutch than there is with the rear brake. The rear brake is a kind of seat-of-your-pants feel, where the transmission is a bit more precise. Especially when you're braking hard and the rear wheel is not really contacting and it's doing this," MacKenzie taps the table with his hand to show a rear wheel that is chattering, "in the corner and losing contact and gaining contact."

The fact that few people use it to capacity helps clutch reliability. Also, because the races are generally only forty-five minutes or so, the clutch is not exposed to long hours or prolonged use. "We never really have clutch problems," continued MacKenzie. "If someone practices a start or something on the warm-up lap you might have a problem for the race, but not in the race itself."

Getting the Power To the Ground

Currently, all Grand Prix bikes use chain drive. Like most of the other components in the drivetrain, chains rarely fail. They will stretch an amazing amount, and the effects can be seen in most close-up shots as the chain begins to hang. The chain is made out of high strength, high quality steel.

"Generally we put a new chain on for the morning warm up," explained MacKenzie. "That will stretch a couple of inches—between 1-1/2in and 2in. That may be an extreme figure for one warm-up, but it's not less than 1in for sure. Once it gets to it's maximum, it won't stretch any more. Then you can use that chain for the race and on into the next practice. We change them often so that they don't break. Pretty much everyone changes it every race just so there won't be a potential problem.

"I remember in club racing they used to break because they had them on there pretty much forever," MacKenzie said with a smile. "And guys shook chains off and broke chains. But normally before that, you'd be chewing up sprockets because there would be so much wear in [the chain]. It would be bad economically to keep the chain on for too long."

The sprockets, combined with the transmission gearing, conspire to make an almost infinite selection of gearing possibilities. And there seems to be no rule that simplifies the matter. The selection of the sprockets and the gears are left to the engineers who can do what they please on any given weekend. Aided by statistics from previous races, the sprockets are matched based on what feels good.

"What we have is either a computer or we have charts, and we start off with the gearbox," said MacKenzie. "With the six speeds, we can change any ratio to any speed we want depending on the rpm and

Clutch discs are cleaned with a solution that prolongs their life. They are sintered metal that will slip when there is little friction, but will grab well when a decent amount of force is applied via the clutch springs.

if we want more in a corner or less in a corner. Then we decide if we want to go faster or slower in each gear.

"We punch all that into a computer and then that will give us the speeds. To get the speeds requires a combination of changing the primary gears inside the engine and the sprockets and gear ratios. There are three things: the actual gear ratios in the gearbox, the primary ratios—that's the one that the crankshaft drives the gearbox with—and then the sprockets. It almost gives you an infinite combination of gear ratios. I mean, we've got something like a million-and-a-half combinations. So that means that you have enough to find the gear combination you want.

"One guideline is that you don't want the spaces too big. Like you don't want first and second really close and then a big jump to third, because it will just kill the power. You kind of try to keep the spacing even, maybe 25 or 30kph maximum between each gear so that it's smooth in each gear—maybe fifteen hundred revs."

A Team Roberts crewman reassembles a clutch after Saturday's practice session. The clutch discs hang on the footpeg.

CHAPTER **8** SAFETY

What separates Grand Prix racers from national caliber road racers is more than just being able to go fast. It's a mindset. These men have a technical understanding of the dynamics of a motorcycle at speed and can fling it around a narrow band of asphalt with precision, sliding the machines deliberately at 160mph. They understand, too, that in a car there is some protection from the hard asphalt and concrete walls. On a bike there is nothing.

"Racing a motorcycle at that level doesn't last very long," explained Kenny Roberts. "Once you get to where you can race for a World Championship, you don't have a long career. Not like car racing. You come so close, and you push the envelope so may times, and it bites you X amount of times a year and pretty soon you get a little leery. You lay your butt on the line so much that it tends to almost overwhelm you that you've got to get up and do it again."

Roberts should know. Heading into the 1993 Italian Grand Prix his rider, Wayne Rainey, was separated by only seven points from Kevin Schwantz for the title. On the ninth lap, Rainey fell, slid for 100ft or so, and then was slammed in the back by his pursuing motorcycle. His sixth vertebra was broken, leaving Rainey paralyzed from the waist down.

Schwantz had been poised to capture the championship in previous years as well. And each time Schwantz had crashed himself out of contention. On several occasions throughout his career, he has put himself out of a race, not to mention the times others have taken him out or the occasions where he has laid the bike down in practice. Like most of the front runners, Schwantz has suffered injuries, but he continues riding his bike as fast and as aggressively as ever. Even Rainey, who, as a soft-spoken, atypical racer had a reputation for staying out of trouble, crashed at the end of 1991, but managed to return in 1992 to win the championship.

Three-time World Champion Wayne Rainey's fall at the 1993 Italian Grand Prix broke his sixth vertebra and left him paralyzed from the waist down. Rainey, known as both an intelligent and careful rider, was poised to take yet another championship before the crash. The accident, which occurred at the peak of the amiable Californian's career, stunned the Grand Prix fraternity.

For his part, Schwantz will take the World Championship anyway it comes. For Schwantz, who tempered his crash-or-win style in 1993 in order to be in the hunt for the championship, the year-end honor

Unlike auto racers who have a cocoon to protect them, Grand Prix bike racers are limited to the armor they carry with them. Physical training is essential. "You don't have to be strong as in physically strong, but I think you need to be able to maintain a certain amount of fitness because the races are so long," Kevin Schwantz said. "You're forever using your forearms and your hands, the middle section of your body, and your legs. There's a constant pounding on it *lap after lap. It just kind of works on it. If you're not conditioned you can be as fit as you like—but if you really ride the bike you'll be hurting after about five laps and you'll need to stop because the muscles you use on a race bike are totally different from anything else. Your arms are going to be pumped up and you may not have the strength to hang on to the thing."*

was bittersweet. Nineteen ninety-three was his most consistent year since he started Grand Prix racing full-time in 1988, and he was the odds-on title favorite even before Rainey's crash.

"I'm going to think back happily on [that] year. Because I feel like we worked really hard. We earned it," Schwantz said days after Rainey's accident. "We didn't just collect it. There were lots of other guys that could have won the championship as well, but we performed well all season. We were consistent. In a way, I feel we deserved to win the championship." Then he paused a long moment. "Of course I'll look back on it

at the same time and sure wish that hadn't happened to Wayne. I'd rather have been second in the World Championship and still have him racing."

As auto racing continues getting safer and safer, motorcycle racing stays relatively static in its record for avoiding rider injury. In an automobile, the designer can address things like cockpit strength, footbox area, carbon fiber density, roll bar clearance, and overall structure of the car. None of that applies to a motorcycle. There is no safety harness. Strapping the rider onto the machine would only make him more vulnerable. As for design, there is little that can be done to enhance

Jim Filice, 1993 AMA 250cc champion and winner of the inaugural 250cc U.S. Grand Prix, was chosen to ride a 250cc Suzuki for the Grand Prix the weekend following Rainey's crash. It was Rainey who helped Filice get a ride for the U.S.G.P. "Racers work hard and struggle to make it," he said. "I guess it makes us strong inside. We have something in us—the will to win, courage."

The FIM continually strives to improve safety. Strict rules exist for circuits, and seemingly insignificant details—like rumble strip design, runoff area, apron material, wall placement and height—are all carefully inspected by the FIM prior to the running of a race. Often the tracks must make major changes or lose the event.

safety. Anything that aids in safety is directly related to performance and how smoothly the motorcycle performs its duties. If it shakes its head or bucks, if the tires are too sticky or unpredictable , if the power band is too narrow, the rider's life is literally in danger.

If a race car driver exceeds the limits of the machine, he loses time, puts a wheel off the road, spins, or hits a wall, most of which are not so potentially costly. The likelihood that a racing car driver will die is remote. That is not the case in Grand Prix motorcycle racing. If the rider is lucky, all that will happen in the event of a crash is some bruises and soreness. Unfortunately, there are few harmless crashes in motorcycle racing. The only acceptable way to deal with it is to deny the presence of danger—or at least ignore its consequences.

"What you think of as you're falling is, I hope this doesn't hurt . . ." 1994 World Champion Mick Doohan said with a laugh. "You can't really prepare yourself for the fall, really. You're only really imagining what you're going to do. You just relax more than anything. You know you're going off. There's nothing you can do about it. You just basically want to get it over and done with and hopefully walk away. But it's part of the job, I guess. Everyone falls down, whether it's car racing, horse racing, or whatever. Unfortunately, you fall off the things or crash them, and it's something that's out of your hands as well. It's a scary thing right before it's about to happen.

"As it happens, you're not really thinking about it, but it is better to slide and get away from the machine and also stay away from walls and other objects. The problem is when the bike lands on you or you hit it.

"Just about every time, you feel you are just about able to save it, and then it just kicks you that little bit harder and throws you off. You feel you can save it all the way until the end, and then it happens."

In 1992, for example, Doohan crashed heavily at Holland's Assen circuit. A splintered leg required special treatment, which reportedly was performed incorrectly by the local doctor. By the end of the first evening Doohan's leg had become infected. Doctor Costa, the official FIM/Grand Prix doctor, checked the young Australian and found the leg dangerously close to gangrene—and possibly amputation. The cast was stripped off and the leg treated properly. Later Doohan's legs

The courses are lined with hay bales—a safety standard since the 1930s. The hay bales help stop a rider from flying into a hard object such as an Armco barrier, concrete wall, or other non pliable object. Perhaps a bit primitive, but effective nevertheless.

had to be sewn together to induce healing of the leg—all procedures that could have been avoided. Doohan credits Doctor Costa with saving his legs.

Doohan's leg eventually healed, but now he has no muscle tissue in his right foot. When he rests for too long, the foot draws together like a fist. His ankle has been fused solid, and he walks with a severe limp, which still hampers his riding enough that he slows his motorcycle with a specially rigged thumb-brake. Hobbled as he is, his peers say his riding is as good as ever.

Although crashing will always be a part of the sport—unacceptable hazards are constantly being eliminated, changed, or made more acceptable. But like any large entity, the Grand Prix establishment can move slowly when tradition is concerned. It took a boycott of the Isle Of Man TT by Giacomo Agostini for the FIM to wake up to the dangers of that circuit. Occasionally, it takes a concerted effort from the riders to provoke changes at a circuit, though riders generally remain mute about safety, accepting the hazard of their chosen profession.

"It's life," Doohan said about the crash. "We can't change it. I look at it like I realized it could happen. In fact, it's almost self-inflicted. You don't really accept it, but I've learned to overcome it. I can ride the bike as good as before, and I don't want to sit second. That's just the way I am.

"Anyone who's out there—anyone who's in front—understands the risks or they wouldn't have gotten into it in the first place. I know why I crashed, and all I wanted to do was get back on the bike and get out there and not repeat the mistake."

Wayne Gardner, at great personal sacrifice, has been critical of circuits on the Grand Prix World Championship trail. Not only does it interfere with his concentration on the track, it also affects his reputation with the public. Nevertheless, he felt it important enough for other riders to protest wherever there was a lack of safety.

"The sport is growing and it's making pace in the media world, and I just want to see it continue. If somebody gets injured or killed, then the sport goes back a peg. I don't want to see anybody injured or killed. If we don't bother with it who's going to bother? The organizers don't give a shit. If I don't complain, then nothing gets done. The organizer thinks everyone's happy, and you get the same the next season. If you suggest it, they don't change it. But if you moan and complain, then something gets done. I realize there is a budget and you have to work within the budget, but things can be done."

The FIM continues striving toward safety. But unlike Winston Cup, IndyCar, or Formula One car racing, the sanctioning body has a hard time legislating safety. It's not as easy as extending the space in the footbox or increasing the roll bar diameter. The best the FIM can hope to do is increase track safety to a point where the riders only have to worry about hitting one another—which is pretty much what they worry about now. High-side accidents generally happen individually, with no outside provocation, and the rider who experiences it is quite often injured. There seems to be no way to resolve that issue.

Strict rules exist for circuits. Seemingly insignificant details (for auto racers, that is) like rumble strip design, runoff area, apron material, wall placement and height, are all inspected by the FIM prior to the running of a Grand Prix. Often the tracks must make major changes or lose the event. The courses are lined with hay bails—a safety standard since the 1930s. The hay bails help to stop a rider from flying into a hard object such as an Armco barrier, concrete wall, or other nonpliable object. Perhaps a bit primitive, but effective nevertheless. There are also new and better ideas in the works.

In 1991, a company named Airfence designed and built a system that has proven to be a life-saving track feature. Made of a nylon/plastic material, the fences are actually inflated and provide a soft place for riders to land should they crash and be flung to the outside of the track.

The fence is formed in 26ft long sections each 4ft high by 4ft wide. Each section consists of four separate

Helmet Evolution

A motorcycle helmet acts as an individual egg carton of sorts, protecting the most fragile organ in the body—the brain—from trauma. Current helmets do an admirable job of accomplishing such protection. They have also obviously come a great distance. As pundits will remark, they still have a ways to go.

The first helmets were poorly designed, poorly fitted, cumbersome contraptions that offered little protection. Postwar era designs were what we would call fiberboard, or particle board—kind of wood pulp—which was covered in leather. They evolved into hard hat-looking contraptions suspended by a series of strings and rivets which kept the helmet off the skull itself. In the event of an accident, the helmet would crush, the internal hammock setup would collapse immediately, and the head would smack the fiberboard—or road, depending on how well the leather kept the thing together. The only good the early helmets did was to somewhat reduce road rash, though the face was still vulnerable. They were, in fact, virtually useless.

The first breakthrough for head protection came, ironically, following the head injury death of an car racer named Pete Snell. In 1957, Snell's friends began a small foundation that would inevitably alter helmet safety standards. Tests were conducted with the fiberboard helmets and, of course, none were adequate.

Directed in part by the Snell Foundation, changes to the outer shell were quick to appear, but more significant were the changes made to the helmet interior. Instead of the ridiculous suspended liners, the foundation suggested something that would absorb the impact. The first polystyrene foam was used in the late-1950s and early-1960s and has remained as the best basic idea in helmet design.

The first designs were just adequate. The shells were thin, the liners far from what would be considered sufficient by today's standards. Basically, helmets were still large caps. They were far better than the fiberboard, but still provided no protection for the face and moved laterally on impact. It was not until the mid-1960s that helmets covered the whole head except the face, and not until the early-1970s that full-face helmets appeared.

The impact-absorbing properties have been continually enhanced over time. The liners have become thicker and provide more protection. The comfort aspects have improved as well, with layers of padding making helmets much more plush inside. But as the protective and additional padding have increased in thickness, the shell has had to grow to accommodate more inner filler. That in turn makes the shell heavier, and the helmet more ponderous overall. Grand Prix riders need to be comfortable for an hour's worth of racing. A heavy helmet adds another variable to an already difficult job. And in the event of a crash, a heavier helmet tends to snap the riders head in the direction he's traveling when he hits the pavement or gravel.

Shells have been lightened slightly with the use of things like Dynema, Kevlar, and carbon fiber which are typically mixed with the ubiquitous fiberglass. The strength gained from the use of such materials makes up for the slight weight penalties, and the overall integrity of the shell remains the same. Essentially, however, the ingredients of the helmet have not changed since the breakthroughs of the early-1960s.

The liner is designed to spread the load of the impact over the head, reducing the shock to any one location. Once the helmet has impacted anything, it is useless; the shell is essentially designed to self-destruct on impact. The internal Styrofoam is also useless after impact because it will not spring back to its original shape. The helmet must be discarded.

So what can be improved?

Standards imposed by safety-conscious government agencies have in some cases conspired to keep the evolution of the helmet—even the racing helmet—from advancing. Often, head injuries occur as the brain moves about violently in the skull when the helmet impacts a solid object.

The Snell Foundation feels it has the remedy for the situation but has been unable evoke change. Snell believes the government specification regarding the "duration" of G-forces is set at an unacceptably low level, meaning that the helmet liner is made with inherent weakness. If the Styrofoam liner is essential because it transmits the impact throughout, then the shell, which takes the impact by shattering, clearly works on a different theory. Snell feels the best helmets will inevitably be ones with even stiffer shells—ones which take the shock of the hit and lessen it by transmitting the shock through the entire shell, rather than by the fracture of the shell material. Tests have shown that the human head reacts to trauma in ways different from what was originally accepted. Evolution continues.

---

air chambers so that should several riders fall, the bag will deflate completely upon impact, but will allow four separate impacts. Internally, the bag features a pressurized frame which collapses upon impact, but is able to be re-inflated within five seconds.

In addition, wide sand pits and huge grass apron areas have been the design standards for new or refurbished tracks in recent years. Most new racetracks are now slower and contain far more soft landing zones than in the past.

Opposite page, *"You never think you're going to crash until it actually happens,"* Mick Doohan commented. *"Mugello was the same thing. I never thought I was going to crash. There, it was just a matter of staying on the road. It was just a matter of staying on the black stuff instead of on the dirt and also making sure I wasn't going to fall off any further. I've been in a position before where I've fallen off the inside of a bike that far, and I couldn't get back on. So I just had to let it go. I was just hoping I could get myself back on the bike and I was just fortunate enough that I could."*

Some riders, like Alex Barros, hang off the bike so far that their elbows rub the tarmac. The rider is always a potential projectile. His protection, other than leathers, helmet, and some body armor, is his own semi-absorbent body. Unfortunately, the human body is a bit like carbon fiber: the structure (the skeleton) is meant to splinter and crack to absorb the impact.

But as machine performance and speeds climb, the danger remains as great as ever. At Hockenheim, Shinichi Itoh surpassed the 200mph mark on his Honda. Impressive, indeed, but it leaves no room for error—in either track design or in the saddle.

Personal Protection

Helmets and leathers remain the primary personal safety equipment the riders employ.

Leathers are exactly that—leather overalls that protect a rider in the event of a fall. Obviously, the protection only goes so far. If you wear heavy leather gloves while using a chainsaw, they help keep your hands from being scratched by the branches and wood chips, but they would do little should you try to grab a running saw's chain. Same with riders' leathers. They do well for the sliding and falling part, but no matter what, there is little protection from a fall on pavement at 200mph.

Articulated spine protectors cover the back from the base of the neck to the seat of the pants. Designed similarly to the human spine, each piece flexes with the back and acts as a shield should the rider slide back-first into something or be hit by a motorcycle. The protector's "vertebrae" are often made of fiberglass with polystyrene filling or hard plastic. Versions are also made that cover the hips as well as the rider's kidneys. In addition, all riders wear shoulder, elbow, and knee pads, which all but eliminate "road rash." For the most part—especially in a low-side crash—the leathers and pads will allow a rider to walk away from a low-speed slide. The leathers offer little resistance, letting the rider slide for a spell, allowing him to scrub off speed. Inside the suit, the rider is usually relatively unscathed.

In a high-speed crash, the leathers will still help avoid the scrapes and initial bruises, but like throwing a sack of potatoes out of a moving truck, the bag may not break, but there are going to be some mashed potatoes. In that vane, there are some suits (which to date have not made a splash in Grand Prix racing) made of Kevlar—the same material used in bullet-proof vests. The Kevlar suits are lighter and perhaps a little stronger, but so far have proven to offer the same

Wayne Gardner crashed at the U.S. Grand Prix in 1989 while making an aggressive pass. The broken leg he suffered kept him out of the championship for several races, and when he returned, he re-injured the leg in yet another crash. Shaken and lacking confidence at year's end, Gardner trained on dirt bikes in Australia during the off-season. By the beginning of the new season, Gardner had sufficiently blocked the pain out of his mind. He tried the same pass in the same corner and suffered the same result: a crash. This time he emerged without injury. Such is the mindset of a successful Grand Prix rider.

Mick Doohan was within a few points of winning the1992 World Championship when he crashed in Holland. The accident certainly cost him the championship, and it nearly cost him a leg. It was nearly a full season before he performed as he had prior to the crash.

level of protection as leathers in a high-speed crash. For now, Grand Prix riders stick with leather.

A set of leathers will last several years on a club racer—but Grand Prix riders change them as often as every race. They will have different leathers for different circuits as well. Some have more ventilation and some less. As for protection, they are essentially the same suits. They cost around $1,500.

Although debatable, leathers probably don't actually save lives. A helmets on the other hand is the one safety item that absolutely keeps riders from losing their lives. There are helmets from several manufacturers—most overseas—used in Grand Prix racing today.

The best helmets are made of fiberglass. Inside the helmet is what you'd expect to see in an ordinary street bike helmet available at any motorcycle shop: Styrofoam. Although manufacturers have experimented with liquid and air bladders, there have been no breakthroughs with those other materials.

If the head is injured, it is usually from a severe concussion caused when the brain moves about in the skull. That remains the biggest challenge for helmet makers; to somehow protect the *inside* of the skull.

Crash Philosophy

Another problem for designers and safety crusaders is the racers' riding techniques—which can be downright terrifying to watch. When riders like Alex Barros hang off the bike so far that their elbows rub the tarmac, there is clearly nothing officials or safety equipment designers can do by cross their fingers. A rider

This soft barrier, produced by a company called Airfence, is one alternative to hay bails. Made of a nylon/plastic material, the fence is actually inflated and provides a soft place for riders to land should they crash and be flung to the outside of the track.

cannot be strapped to his bike; he remains a potential projectile. His protection, other than the leathers and helmet, is his own semi-absorbant body. Unfortunately, the human body is a bit like carbon fiber: The structure (the skeleton) is meant to splinter and crack to absorb the impact.

In that case, the best the FIM can hope for is that the attending doctors and medical staff are adept technicians as well as analysts. To ensure that, the FIM has Doctor Costa check all aspects of the medical facilities at each track from evacuation and safety roads to the actual administration of treatment and medicine. If the Costa doesn't perform the treatment himself, he is in close communication with the attending physicians.

Most riders don't like to discuss injuries or crashing—they prefer not to dwell on the danger. It's just part of the job, and the risks of racing. Injuries cannot be eliminated, so why think about it? But there is a process. There is a way a rider stays mentally fit, and it happens almost every time he crashes—even if it's only for a fraction of a second. The thought of giving up.

The body is apparently fine, although the leathers and the helmet will be tossed at the first opportunity. Fortunately, riders' memories are very short when it comes to accidents. This racer was back up and racing less than two hours later.

The rider was uninjured, but the bike was not so lucky. It can all be fixed, of course, but the adjustments were likely off for the next practice. At least the rider was in one piece.

"Sure, sometimes you think like that. But today it wasn't a situation like that," Doohan said the evening after his tumble. "Sometimes you have a race where you get up, look at the bike, get back on, and continue on. Certainly sometimes when you knock the stuffing out of yourself you kind of feel like, What are you doing this for? But you sit down for a few days, and you always want to hop back on the bike again. It's almost like a drug. You always want to keep doing it."

The best safety, however, is not done via other people or with objects, but with the rider and his conditioning—both mental and physical. As we've been discussing, the mental conditioning is at work constantly, trying to minimize the danger, trying not to dwell on the possibilities. But the more physically fit a rider is, the better he can maneuver the bike in times of trouble. It is a punishing sport, and the rider with the level of stamina needed to control a bucking motorcycle after forty-five minutes of racing is the one who can still hang on after the bike tries to toss him nearly an hour after the start of the race.

"You don't have to be strong as in physically strong, but I think you need to be able to maintain a certain amount of fitness, because the races last such a long period," Kevin Schwantz said. "You're forever using your forearms and your hands and the middle section of your body and your legs. But it's not so much or so strong, its just that there's a constant pounding lap after lap. It just kind of works on you. If you're not conditioned—you can be as fit as you like—but if you really ride the bike, you'll be hurting after about five laps, and you'll need to stop because the muscles you use on a race bike are totally different from anything else. Your arms are going to be pumped up and you may not have the strength to hang on to the thing."

Said Doohan, "Some guys need [massage therapy] but I think it's psychological as much as anything. I never used to have it [massage] when I first started racing. I don't think I need it just yet. Sure, when I'm hurting sometimes it's good to have a massage, but it actually takes my concentration away from what I'm doing."

And in the end, concentration is really the only safeguard a rider can rely on.

This accident resulted in only minor injuries for the rider. The bike traveled quite a distance without its rider and was considerably worse for the journey.

CHAPTER 9 AERODYNAMICS

Compared to an open-wheeled formula car, a racing motorcycle has a superior aerodynamic form. The car has two huge areas which are exposed to the air and which cannot be eliminated: the tires. Regulations in all open-wheeled classes say that the front wheels must be unencumbered by bodywork and exposed. The car's open cockpit, radiator areas, and rear wheels also slow the car. So unless you discuss prototype

This 1952 250cc NSU Rennmax featured a tail fairing.
Mick Walker collection

Full dustbin fairings were commonplace until they were banned in the late 1950s. Here we see the 250cc dueling dustbins of Moto Guzzi-mounted Enrico Lorenzetti (no. 21) and MV rider Carlo Ubbiali (no. 23) at Monza in 1956. Mick Walker collection

sports cars, the major classes of auto racers are not very slippery through the air.

The motorcycle, on the other hand is a fairly slippery form. Formula cars have become slimmer and taller over the past few seasons, but motorcycles have always had that figure. Having two fewer wheels is another clear aerodynamic advantage the bike possesses. At the same time, it is very thin, and relatively streamlined. Even stripped of all its fairings, a motorcycle will be much more aerodynamically efficient than a modern Formula car.

Motorcycles, in fact, have at various times throughout history actually been faster than automobiles. In 1907, for example, aviation pioneer, Glen Curtis, fixed a large aircraft engine within a crude fairing on a motorcycle and ran 136mph. German manufacturer NSU ran a 500cc machine to 210mph at the Bonneville Salt Flats in 1956; they also helped bring aerodynamics into Grand Prix racing. In fact, some of the fastest racers on the Bonneville Salt Flats are motorcycle-type vehicles. Less wheels equals a more slippery package, no matter how advanced the technology.

"Formula One cars are not very slippery at all. It's possible to make a motorcycle fairly good," said John Mockett, renowned motorcycle aerodynamicist and occasional consultant to Kenny Roberts. "If you run a motorcycle in the wind tunnel without a fairing you will get quite a bit of lift on the front end. If you think about it, the bike actually pivots around the rear axle. So if you imagine the rider as a kind of sail, the wind will push him backwards and rotate the motorcycle around the rear axle. It will give you lift on the front. So if you then reduce the drag and make the sail smaller on the top, you will reduce the rotation around the rear axle by making the shape more efficient. If you put a fairing on—unless it's a ridiculous shape—you will tend to reduce the lift on the front wheel. But you don't ever actually get rid of it."

"I made a fairing for John Kocinski when he first came over here and rode a 250," said John Mockett, occasional aerodynamic consultant to Team Roberts. "Whenever you read a bloody magazine article they always bring out the photos of John's bike and say that she's the most aerodynamic motorcycle of the last few years."

So motorcycles are inherently more stable in the air stream and more efficient. But where the automobile can be changed to direct air in a positive way, changes to the bike are limited. In the early 1950s, racer and car builder Jim Hall found that if he played with the bodywork of his sports car he could not only minimize lift, but he could produce downforce. Since that time, cars have pushed the aerodynamic envelope in search of grip. Ever bigger or more efficient wings help push the race car down on the track at more than twice the weight of the car. But wings don't work on motorcycles. No motorcycle fairings will actually achieve a negative lift on the front end. The faster they go the lighter the front gets.

"The only way you can start to do that [get downforce] is to do what the car people have done, which is inappropriate for a motorcycle, which is to put anti-lift devices on them like wings," said Mockett. "They are inappropriate because motorcycles lean over as they go around corners. You would get enormous instability and strange things would start to happen—think about the yaw and that sort of thing. It might work, but it's not something that I would encourage anybody to spend lots of money investigating," he said with a laugh. "And of course, it would make them slower too."

The aerodynamic returns on an automobile are created by losses in other areas. The cars sometimes have 700-plus horsepower to work with. A loss of 100hp due to drag created as the car produces an extra 300lb of downforce can be justified. Cornering speed will rise, and outright lap speed will increase even though the car's top speed will probably decline. Consider Indy: In 1972, when the entire field began using wings to increase performance, the overall lap times were almost 20mph higher than the previous year. Cornering speeds were nearly fifteen percent

Utilitarian as well as uncomfortable. As part of the rear bodywork—and consequently an aerodynamic concern—it behooves designers to mold the fiberglass in a single piece.

faster, but the straightaway speeds were down more than 10mph. Just as the designers had created downforce for cornering, they had also created drag that manifested itself in slower straightaway speeds.

On a bike, the cornering speed cannot be enhanced because as the bike leans into the turn, a fixed wing would push sideways rather than vertically. The gains would only come when the bike was upright and traveling in a straight line—in which case the terminal speed would be adversely affected. If the rear wheels could remain in the vertical position and not lean over as the bike did, then vertical downforce might be useful. But for now, no such configurations exist. To sum up: aerodynamics on bikes is limited to reducing drag. The more blade-like the motorcycle can be built, the better it will perform. The rider has to do the rest.

"Basically you're creating an enormous amount of wake," explained Mockett. "The best aerodynamic fairing or aerodynamic devices reduce the amount of wake that you leave behind. You create a sort of a low pressure area behind you and the air sort of rushes back into that. So the smaller the wake you create the more efficient the form will be. If you think of a razor blade, for example, it doesn't create a large wake unless it is at a slight yaw, and then it will create an enormous wake. But absolutely edge-on, it will tend to cut through the air, as opposed to a brick which will cre-

Opposite page, already a fairly slippery form, the motorcycle's aerodynamics have changed little. As formula cars have become slimmer and taller over the past few seasons, motorcycles have remained largely unchanged. Even stripped of all its fairings, a motorcycle will still be much more aerodynamically efficient than a modern Formula car.

The budgets in motorcycle racing are tiny compared to the budgets in car racing. "Either you find a very small outfit who is prepared to spend a lot of time on [aerodynamics] out of interest value, or you find a huge team with a massive budget and a separate test team who are able to do it," says John Mockett. Cagiva has the resources to experiment.

ate a tremendous disturbance. So what you have to do is not only create a flow of the air as it hits the front of the bike and take it where you want to take it, but you also have to think about trying to put it back together again after it has gotten past the bike and rider. And most motorcycles don't take that into account."

The classic race motorcycle riding position is that of a fetal-like crouch, hidden from the air by the screen, and more or less doubled over and laying as flat on the fuel tank as possible. The position makes perfect sense; riders are trying to maximize the bike's inherent ability to cut through the air. The less surface they present to the air stream the better.

The other area of importance to aerodynamicists is the ability to move the air through the motorcycle. The radiators need air for cooling, and the carburetors need air to mix with fuel for combustion. Each is an indirect function of aerodynamics. Ideally, what designers want, said Mockett, is "just a complete blob on the

Aerodynamics on bikes is limited to reducing drag. It is also, however, important to make sure the engine is given sufficient air both for intake as well as cooling. Note the 1993 Cagiva's ducting.

For 1994, the Cagiva's nose was more blunt with aerodynamics forsaken in favor of improved airflow.

front with no holes in it at all." But if the holes must exist—and they must—the best a designer can hope for is a clean, efficient air channel. At one time, the goal was as straightforward as that. But through history, a few things have abated that style.

"If you watch the classic style of a 125 rider he will try to stay as small as possible behind the fairing for as long as possible," Mockett said. "The 500 riders have far less corner speed and they have to kind of jump around to control it. So they're sitting up a lot more, and they're performing gymnastics even down the straight sometimes to keep the bike straight and level. So they tend to be less interested in maintaining an aerodynamic profile than the riders who've only got 40hp to play with."

The classic style was no match for the Kenny Roberts' bike-bullying style. Finesse had its place in the lighter, less powerful classes, but in the world of the big 500s, the fast guys were powering the machines out of the corners. Roberts still stayed low and thin on the fast sections, but entering a corner style went out the window. No matter how much the designers were able to enhance the airflow, the riders were not going to be passive passengers on the machine. So the aerodynamics became secondary to engine output, which was the great equalizer for an ill-handling bike.

In the final analysis, the look of the racing motorcycle hasn't changed much in the past three decades. "It's an incomplete form," said Mockett. "You'll always run up against that. The only way around that is to be very radical. You can only sort of reckon to make so much improvement. It's like any job: The first 90 percent is quite easy, and the next ten percent after that is as hard as the first 90 percent. So as things get better you make smaller and smaller gains. "

Looking back on the designs, the early fairings looked rather unsightly, ungainly, and plain low-tech. But, explains Mockett, they were exactly the opposite. The bigger more rounded look was much more efficient than today's design.

"The thing about the Yamahas is that their fairings were always very angular. Even kind of darty. That isn't a very good shape for a slow moving body. It wants to be much more bulbous and zeppelin-like. But of course, that wasn't really very fashionable. The mid-1970s stuff . . . some of that stuff was actually pretty good.

"The reason they prohibited people from covering the front wheel is that they started going really fast. The old Moto Guzzis were trouncing along at a really high rate of speed. And of course the tires were terrible—I mean, tiny, tiny tires. Compared to today's tires they were very low-technology bits of stuff, and they were throwing treads and doing all kinds of things like that. I mean, they had tubes then and spoked wheels. Really, they started to go much too fast. They also said that they were having problems with side winds, but really I think that the chassis wasn't up to the speeds that they were able to achieve. So

Spontaneous aerodynamic modification. Here an Aprilia crewman works on an internal airflow box with a small band sander.

[the FIM] said, 'Let's slow them down.' They did, figuring, Well, those 'dustbins' are too dangerous, so we're going to limit them to a dolphin style fairing, which was a terrible sort of retrograde step.

"The NSU Flying Hammock looks like what I'd call a record breaker. It looks like a cigar, with the guy sitting way in back. And when you look at the speeds that thing achieved, it's absolutely incredible. What's more incredible was that it was actually usable; its wasn't just a straight-line vehicle. NSU actually built a Grand Prix team for that year. What happened was that [Gustav] Baum crashed one at the Nurburgring and was killed. They canceled the whole program because of it. In actuality, the crash wasn't from the vehicle; it was because he was a designer not a rider, and he made a mistake. If that hadn't happened—if that bloke hadn't been killed—I think the whole sport of motorcycle racing would be very different now. Those things were very good."

Clearly much of the stagnancy of the aerodynamic form came from the sanctioning body's attempts to keep racers from killing themselves. Although a noble cause, the result has become a source of irritation to designers who see the obvious way of making the machine more stable.

"The interesting problem is that the regulations stop you from doing what is very logical. You can reduce the drag coefficient enormously if you bring the fairing down over the front wheel—which you're not allowed to do. And in fact, the fairings are much lower in that area then they are meant to be because [the FIM] made a mistake a long time ago by saying that from the side the wheel should be exposed. But what they meant to say was that the *tire* should be exposed. But they didn't say that. So people have taken advantage of that to mask the front wheel to a certain extent. And if you take a look at that over the past few years, you can see the shrouding of the front wheel becoming more and more apparent to [a point] where a lot of the machines were illegal. The distance forward—the nose—was only supposed to be as far forward as the font axle. And then they allowed you to go 50mm farther forward because everybody was, and then they amended that to 100mm farther forward. So the noses of the fairing are gradually moving farther and farther forward. And when you actually do that—as you improve the efficiency of the fairing in terms of drag—you reduce the tendency for the front end to lift as well."

The regulations that kept a more complete fairing from being designed and utilized, paired with the fact that some riders preferred to hang off the bike

A fully faired NSU land-speed racer. The entire bike was enclosed for optimum aerodynamic performance. Obviously, full enclosure would not work for a road racer, but it's interesting to see what designers were experimenting with in the 1950s.

rather than hide behind the fairing, helped to stunt the growth of aero engineering. And in the era where riders will switch teams at a moment's notice, the only thing the designers know will be a constant is power. The long-term plan—regardless of whether or not a team has a World-Championship-caliber rider to build a bike around—is to keep making the motorcycle more powerful. The aerodynamic form becomes an afterthought to keep the engine hidden from both the competition's eyes and from the wind—and certainly the former is perceived as more critical than the latter.

"The problems that you get into when you get into [aerodynamic development] is that the teams have spent an enormous amount of effort and time in tire testing and suspension testing, and if you produce something that effects that suspension setting or the tire they don't want to take another variable on board," said Mockett. "So sometimes you produce a fairing that works pretty well, but does effect the handling and they're never prepared. The budgets in motorcycle racing are tiny compared to the budgets in car racing. And they just can't take that sort of thing on board as well. Either you find a very small outfit that is prepared to spend a lot of time on that [aerodynamics] sort of out of interest, or you find a huge team with a massive budget and a separate test team. But I'm afraid that any sort of aerodynamic advances will not be made in motorcycling overnight. They will correct them, they will get better, but not that quickly."

Nevertheless, it is an area with tremendous po-

The motorcycle is a fairly aerodynamic form as it is, and a normal, run-of-the-mill fairing design will simply enhance the package. This one, produced by Elf in 1978, doesn't appear so efficient in hindsight.

tential. "Obviously we've gotten to the point where power isn't really making better lap times," Kenny Roberts said. "I think that they've got to improve aerodynamically. I think that because we've got an excess of power, we need to start pushing some wind in the right direction to make it go around the corners faster. That's really the next step, and of course if you can match that with some reduction of weight, I think you'll really see the motorcycle go faster."

At the moment, motorcycles are designed around engines and power output. Most designers feel it would be productive to examine that lopsided focus, and perhaps move the components around on the bikes to take advantage of the aerodynamic benefits. Not just in terms of the outside skin, but also in the way the air is utilized once it enters the motorcycle.

"The problem is that people are quite willing to spend massive amounts of money in the area of engine development," said Mockett, "but they're less keen on putting money into wind tunnels. So it's quite difficult to get people to use them. It's not recognized as being very important by people in the sport. I think it's beginning to catch on now. But I've always had a bit of a problem trying to convince people that it's worth doing." Mockett paused for a moment, "I made a fairing for John Kocinski when he first came over here and rode a 250," he said with a noticeable bit of irritation. "Whenever you read a bloody magazine article they always bring out the photos of John's bike. And they say that she's the most aerodynamic motorcycle of the last few years, but if it's the most aerodynamic motorcycle of the past few years why don't people actually have a look at it. It's crazy."

More often than not, designers like Mockett get frustrated with the progress and find themselves doing something else. With so many variables, the Grand Prix bike will continue to make changes in aerodynamics only as the other components dictate it. The radiator, for example, could be re-located. It is not only a drag on the bike, but there is really no answer to the question of where to exit the air once it travels through the radiator.

"Motorcycle race bikes are styled—they're not designed really. There are improvements. If you look at the location of the inlet ducts for carburetion you'll see that they got better. Stuff like that," he said with a bit of resignation. "But it evolves more than anything else."

Above, *Elf's later ideas were designed around a land-speed record attempt. FIM rules would never allow this design to run in competition. It would, however, be a far cleaner aerodynamic form than any contemporary fairing.*

Left, *Kevin Schwantz's 1994 RGV 500 features a very dart-shaped nose. Notice the intake ducts on each side of the front wheel. Aerodynamics didn't seem to help the performance of the Suzuki, though, which was consistently trounced by Mick Doohan's Honda.*

Opposite page, *seemingly insignificant, the fenders do a great deal to cut through the wind and deflect the air to the sides of the fairing.*

CHAPTER 10 ELECTRONICS

The state-of-the-art Grand Prix motorcycle can be compared to an F-18 Hornet fighter plane. In the Hornet, the pilot—even the best in the world—can't keep up with the airplane. The plane flies faster, reacts quicker, and makes precise maneuvers with far more sensitivity than the pilot could ever control. The airplane has been developed to do things that the pilot could never control. Consequently, the plane has been further developed to change the decisions available to the pilot.

The airplane slows itself down as needed.

Unfortunately, compared to fighters, motorcycles are decidedly low tech. Although they are the cutting edge of earthbound two-wheeled vehicles, there is a difference between something that can go Mach One with thousands of pounds of thrust and something with 180hp that can go from 0 to 60mph in 2sec or less.

The major problem is that bikes have come so far powerwise that only the best of the best can stay on them. In that respect, only a handful of people can ride a bike to the maximum degree and actually use their ability to define the limits of the motorcycle and their ability to deal with it. The problem for those elite few lies in riding at the limit. Should the envelope be expanded much farther, it will be necessary to detune the machine, or at least make it seem detuned to the rider. Like the fighter plane.

If Grand Prix motorcycles have plenty of power already, and there is no reason to continue development in that area until the power which is currently available is usable, then there is really only one area which can be explored to further enhance the bikes' performance: electronics.

For now, however, devices to assist a Grand Prix rider are neither crucial to the operation nor simple to develop. Nor, critics say, would the fans or participants want to see the series evolve to that point. Until the 1994 season, Formula One cars had a variety of devices to assist the driver in his task, including active suspension and traction control. These tended to reduce the required driver skill and allowed the racers to slip and slide the cars at will and when conditions necessitated it. They also made for expensive, somewhat sanitized racing which ultimately led to the devices being banned. The goals of a Grand Prix motorcycle engineer are more related to rider satisfaction than ultimate control through technology.

"There's still a major thing which is to make a rider happy," said Kel Carruthers, 1969 250cc World Champion and now a tuner for major 500 teams. "I think on a Formula One car, for example, you can set the car up perfectly and set the driver in it and tell him to go drive it. With a motorcycle, the rider has a lot more to do with it. Okay, you try to set the bike up as perfectly as you can, but he's got to be happy with it. If he's not happy, he doesn't go fast. Even if the bike's not setup really well, but he's happy with it, he'll still go. The recorder system is really good to assist you when the rider can't make up his mind what he's feeling and what he wants."

So somewhere between hiring a robot to twist the throttle and hang on and hiring a rodeo star to keep the machine under control, a compromise exists that can make the best bike better and the best rider more effective.

As with most contemporary forms of motorsport, the machines carry data acquisition systems. The data is retrieved when the rider returns to the pits. That is as complicated as it gets. There is no telemetry currently, and certainly no way to adjust any settings from the pits. The McLaren Honda Formula One cars had metered fuel injection with "black boxes" to manage the fuel flow and the mix, and electronic ignition that altered timing, via telemetry, depending on the load to the engine, the heat of the engine, and other variables. (Rumors had it that Honda's telemetry was transmitted via satellite back to Japan so that factory engineers

Behind the fairing, the race bike looks as much like a computer as a machine, but Grand Prix motorcycles are hardly as complicated as their four-wheeled counterparts.

Relatively sparse instrumentation—just temperature and tachometer. There is a lap counter, but there are no buttons for brake bias, radio, anti-roll bar, fuel mix, fuel pressure, mileage, or assorted other functions that you might find on a sophisticated Formula One car.

Cagiva monitors the data processed by the computers. In most cases the data can only point out the problem. The solution is much more elusive.

The "black box" is fairly uncomplicated and not usually a source of problems. "We've been using the same basic engineering for the past thirty years," said Bud Aksland, of Marlboro Roberts. In the lower right corner of this shot you can see the data retrieval system mounted to the side of the Marlboro Roberts Yamaha.

"We control ignition timing and things like that," explained Bud Aksland, Kenny Roberts tuner (center facing camera). Any advantage that can come from analysis is used.

could monitor the engine's performance on any given Sunday at any given racetrack on the planet.) The Honda's telemetry made adjustments from the roadside. The driver, unfortunately, had been relegated to a well-paid throttle-pusher (which may account for what looks like a waning interest in the Formula One show).

The HRC motorcycle racer, on the other hand, enjoys almost none of Honda's Formula One technology.

But, as Kel Carruthers said, a bike is different. The rider is an integral part of the machine, controlling the bike completely. But as in the four-wheeled world, (or, more aptly, in the aeronautical world), the rider may have just a bit *too* much to keep track of. What the computer models say about frame stiffness and tire adhesion, unfortunately, does not work in the real world. According to the data, frames are too stiff as they are. According to the riders, frames are still too flexible and need further stiffening.

Because of this gulf between theory and reality, only a few electronic devices have found their way onto Grand Prix motorcycles. These are nowhere near as sophisticated as the black magic available in Formula One in the early 1990s, but the devices allow a bit more control to a pilot who needs as much help as he can get.

"We control ignition timing and things like that," explained Bud Aksland, Kenny Roberts tuner. "It's nothing magic. The production bikes—the TZ Yamahas—have it also. We don't have engine management, per se. The 'black box' is a generic term. We have an ignition system that's in a black box. You can vary certain things, certain parameters—throttle position and this and that—but its not an engine management system. We have separate data acquisitions—

Cranefield F1 Data acquisition systems." The systems are fairly uncomplicated and, unlike IndyCar or Formula One racing, the systems are not a source of problems: "We don't really have any serious problems any more. We've been using the same basic engine for the past thirty years."

That antiquity has been and will continue to be the biggest area of focus as the engineers struggle to make the machine perform better and better. "I think that we are very much at the infancy of the sport," explained *Cycle World* columnist Kevin Cameron. "There's a sophisticated computer in your microwave oven, there's one controlling fuel injection and spark lead in the dash of your car. The reason they are there is that they are the best way of controlling those functions. Mathematical functions and fuzzy logic are just starting to be applied [to GP bikes]. I think the day will come when fuel injection will be [used] too, and carburetors will be regarded as a very peculiar aberration that went on for far too long. Every [manufacturer's] engine uses a different type of carburetor. In fuel injection, that's all handled in software. There would be one or two injectors maybe, but it would all be auto parts quality stuff. You would address the mixture problems with software. Same thing with suspension."

Stator windings differ little from a standard street bike.

The bikes are equipped to handle a system—at least they're gathering data as if there is a system in place. But for adjustments there is no magic system. A major reason for recording data is not that they are able to use the information now, but because they feel they might be able to use it in the future. Dynamic models exist, but they are not yet usable. The data collected today is used to assist in things like choosing gearbox ratios, selecting overall steering geometry, and ensuring that the suspension doesn't bottom. "Honda has something like twenty-seven [data acquisition] channels," Cameron said, "and they aren't us-ing all twenty-seven channels. Most of it is going into a library. They probably have people looking into analysis. They use the data recorders not so they can pinpoint all the chassis points and change them, but so they don't have to ask the rider what's happening."

A variety of sensors are mounted at various points on the motorcycle: one is mounted on the front suspension, as well as one on the rear suspension, to check travel; another monitors the wheels to measure wheel revs and calculate speed; one more checks exhaust temperature; and often there is one for combustion temperatures. Both of these last two are both checked against rpm. Occasionally the bike's lean angle will be monitored. Teams will operate different monitors depending on what equipment they will be using.

"To use engine management control you really have to have fuel injection because you can't control carburetors electronically," said Carruthers. "I think it probably will come. But motorcycle carburetors really are pretty efficient. With the revs that [the bikes] do, the fuel injection is a lot more complicated than on a four-stroke engine. They run about the same revs, but [the two-stroke] injects twice as often. The stuff we've got now is pretty sophisticated, but we don't control it from the pits or anything."

Beyond data retrieval, the electronic gadgetry on the average bike is used to control ignition. Like any good racing engine, detonation will tend to be a problem. Nevertheless, engineers are looking for a hot spark. If the engine does not have a hot spark, it tends to foul plugs. The biggest problem has been the time available to generate the spark. Electronic ignition helps, but the revs make timing very critical. Everything runs on a generator system. There is a battery, but it only exists to get everything going.

"The thing with a racing motorcycle is that the people that more or less did all the development work have been people like myself who are more practical guys rather than theoretical guys," said Carruthers. "It wasn't until electronics came along that they had to employ theoretical guys. Us older guys didn't know anything about it. Electronics—that's a whole new field. In my day, we used magnetos and coil ignition and everything—much the same as cars; they were all fixed ignition timing. Once they got into electronic ignition . . . well, they can vary the ignition timing through the range, and then with two-strokes they have electronic exhaust valves and all that sort of stuff. Now they're really sophisticated. Nowadays you have people that that's all they do."

Besides the data acquisition, probably the most sophisticated areas on the bike have been anti-dive (or semi-active) suspension, the cutoff that kills the engine as the transmission is moved from one gear to the next, and traction control.

The anti-dive is setup via an RS232 plug, and the bike can be preprogrammed in the pits using pre-existing computations to change the damping force. It could also apply logic that told it that the brakes were on, or that a certain speed was achieved. Or it could have been programmed so that at certain times on a certain circuit—which could be done by a roadside trigger or an onboard trigger—that could keep the bike from, say, bottoming out.

Full active suspension, as defined by an early 1990s Formula One car, is a hydraulic system with no springs. It is suspended by a network of rams and actuators which react to the road, keeping the vehicle level—if so programmed. They have existed for jet aircraft since the 1950s. On full-active suspension, when the vehicle hits a bump, the entire suspension compensates for that bump by both moving the wheel and the chassis. Active suspension uses logic and hydraulic power to change the suspension, anti-dive uses logic to actuate it, but it is no longer electrically—or rather electrohydraulically—actuated.

Another major force of electrical innovation that has been fully developed in Formula One is traction control. Different rider styles keep any one traction control model from being universally adopted by all teams. Most riders, for example, want the rear wheel to slip. Nonetheless, all three Japanese manufacturers have experimented with some form of traction control.

"There are different forms of traction control," said Erv Kanemoto. "Some have been by throttle opening—a certain percent power is cut back at a certain area. It could be done by exhaust valve timing or ignition timing to soften the power. All the factories have worked in this area, and they may do it a little bit different. In most cases, it's a combination of the way the exhaust valve opens and maybe connected to the throttle and maybe the ignition's advance/retard curve. They are overriding things in the exhaust valve opening. What they're trying to do is understand, Well, at this throttle opening on an average would this be best? I believe Suzuki uses ignition—they change the timing—in some of the lower gears only. You wouldn't want the thing to happen when you're accelerating straight up. Most of the cases you want it to occur when you are banked over."

Although the systems are far from universal, they are becoming more commonplace. The "droner" engine took the bike forward a great step in terms of rideability. But many believe the area of electronics holds the key to the future of Grand Prix motorcycle racing. Will the next generation yield computer controlled suspension, electronically programmed automatic transmissions, and digitally metered fuel injection?

"I think that something of that sort may be around the corner," said Kanemoto. "A lot of the factories were able to take away a lot of the devices to reduce the power in, lets say, the lower gears, because the engines were allowing better traction and letting the riders get on the power earlier. But now they're back up again to where the power is again coming in fairly hard. So again, they're working in that direction."

Kel Carruthers agrees, "I think the biggest thing in some respects has been electronics." And only time will tell how much bigger it will get.

Data acquisition allows team managers to detect where the technical problems are located. It also allows the rider to see where he can afford to increase braking or power. In the end, the data recorded is used far less than it could be.

CHAPTER 11 RACE DAY

Grand Prix motorcycle racing, like any competitive motor racing series, quantifies the season as if it is a single period of time, with the result—the World Championship—as the only war in a season of sixteen smaller battles. One event blends into another as the upcoming race actually begins as the riders coast into the paddock following the last race. But unlike most championship series, the next race will be in a different country, in a different context.

The blur of colors, sights, sounds, and people who make up the Grand Prix circus pay little attention to the surroundings, however. They are preoccupied, consumed, with the circuit's nuances. They see only statistics and settings. From last year's race, the team will know what to expect to a certain extent. But the focus will be on how this particular track affects the bike. The rider's style will dictate how the machine is prepared.

"The bureaucracy of Grand Prix racing involves the record keeping of technical records," explained veteran American Grand Prix racer and independent team owner, Robert MacLean. "We have not only our own records of suspensions and tires and so forth, but we also have the opportunity to share with say Yamaha France or with Kenny or whoever. So you never come to the track with an absolutely blank sheet of paper. You always come to the track with a starting point. You know what worked for somebody else last year. You know some basic things. That scenario has been greatly enhanced by electronics. Now you have electronic records of what happened on every lap at every corner. On a given weekend, though, you know the track. Either you've raced there or you've tested there. You know the setup and it's just basically getting the thing setup."

The procedure at the beginning of the weekend would generally be to take a peek inside the engine then run the first practice with the engine from the previous race. Unless the bike needs a new crank or something equally extensive, the team won't make

When the tent closes, the secrets are uncovered. Grand Prix racing is still a very secretive sport, more so than almost any four-wheeled series. By season's end, teams are less concerned with sequestering their equipment, since new technology is on the way.

any major changes. Not yet.

The track is usually pretty "green," so the riders don't have a lot of traction. They busy themselves with getting things sorted out in the pits. Actually, they expect the bike to pretty well setup right out of the box. "As a matter of fact," adds MacLean, "there are some

Dustbin Central! Racers push start their machines at the start of the 1957 Italian 500cc GP at Monza. MVs, Gileras, Guzzis, Nortons, and BMWs fill the grid. Mick Walker collection

races where you'd make some basic setup changes while you're packing the bike for the next race. There are some things you do before you even get there."

Free practice starts at 9:00 am on Friday for 250cc bikes and 10:15 for 500s. Practice lasts an hour. The rider wanders out to pit lane where the crew will have wheeled the machine to the front of the garage. As the practice start time draws near, the bike, now tuned and ready, is pushed and bump started. It coughs to life and belches the first cloud of sickly-sweet blue smoke. Moments later another bike is jerked to life, then another, and soon the entire pit lane is a cacophony of rising and falling engine song.

The rider, saddled with confining leathers, pads, and safety gear, walks awkwardly to the machine and straddles the bike. He twists his wrist and the engine note rises. He listens for the song, looks behind him at the exhaust, perhaps down at the machine, and feels the bike pulsing beneath him. When he feels so inspired, he will engage the clutch and move the machine down pit lane, out onto the circuit.

"The first practice, I go out and usually feel the track out and feel the bike out," explained occasional 250 Grand Prix rider and AMA 250 Champion, Jimmy Filice. "I get comfortable. I usually go out slowly. I get myself comfortable and make sure the track is in good shape. I get some corners in my head and get some gears together for each corner. Then we try to get each corner geared right in the gearbox. That goes on all weekend. There are really two practices on Friday and both are untimed. They are both set-ups for the bike with the suspension and the gearbox."

Grand Prix racing is a near-endless series of adjustments culminating in, hopefully, the optimum set-up for the particular track. The wheelbase, the rake, and ride height can all be adjusted. Many other components can also be changed. Teams look to set the bike up with a neutral handling characteristic where the rider can push—or purposely understeer—the front end a little or slide the rear end. Although the setting is never the same at the next track, a neutral-steering machine gives the crew a baseline at every race.

149

Most riders want the bike setup fairly close in qualifying to what they will run in the race. "Say you're three seconds faster in the race than in qualifying," explained Jimmy Filice, 250 GP winner. "There's a good chance you'll crash in the race because you may have made the suspension too soft to handle the corners. You have to be running real close to your qualifying lap times in the race because it just drives off the corners differently." Here Kevin Schwantz pushes his RGV entering a corner.

"Usually with suspension—with ride height—you have to get to speed. Sometimes you'll reach a point where you can't get around the corner as fast as you want to. You might have a front end push where it drives off the corner and it drives off the apex straight, so you might want to lengthen the wheelbase a little bit. You might want it to steer a bit through the "S" corners and get the bike to turn a little easier. There's some give and take to get where you want to be," said Filice.

The degree of pleasantry is related to how well things go with the tuning. If the weekend has started poorly, it will likely end the same way, and faces are sour, tempers short. When things aren't going well, it gets quiet and stress is evident.

"These are incredible three-day energy bursts, and different riders handle the practices differently," explained MacLean. "Some have different agendas— or, at any rate, the team has different agendas for them. The actual technical stuff is done by Mike [Webb, the chief mechanic] and his assistants. Mike is basically working with Niall [MacKenzie] and getting the bike as race ready as possible. Mike has the actual responsibility for preparing the bike. He just tells us what time he'll be ready for dinner. He doesn't worry about when he'll be at the race, or where he's staying, or anything else. And within the mechanical side or the work side, there are certain things that he will delegate to other mechanics. For example, Mike does the gearbox—there's a certain amount of experience and knowledge that goes along with that—and there are certain things for his assistants to do. Then the gopher, as the season goes along, instead of just moving wheels back and forth to Dunlop, he can do other things. Niall's input, obviously, is used. He'll not only visit with Mike, but with Öhlins and talk about what he thinks about the suspension, and he'll go to Dunlop

To the victor goes the spoils. Here Alex Barros collects his trophy. Judging Kevin Schwantz's face, his result was unsatisfactory. It could have been an adjustment that was not made, a component that was ill-selected or just rider error. Or, perhaps, Barros just did everything right.

and talk to them about tires. He ultimately makes the choice of tires. Everything is focused on making sure he has the most competitive machine possible.

"Our goal at the beginning of the year last year was to be reliable and finish every race. But you need a combination of people who can get put through the ringer and get along living in close quarters. It makes the chemistry very good. I don't care what the business is, the people part of it is very important. The team comes first, and the object is to win races and score points."

The prerequisite for the first day will be to establish a time. The rider will have pushed the bike to its limits all morning, but will want to be clean and safe for the first of two qualifying sessions. The qualifying times should be close to what the practice lap times were—often they are identical.

Saturday morning is untimed practice and the teams will want to test for Sunday's race. "Sometimes at this point we might want new pistons for the race,

and we might use that time to break them in," says Filice. "For the timed practice we should basically have a race setup. Maybe the tires are a little soft to allow you to get up to speed a little quicker and put some good times in without worrying about making it last for thirty laps. Sometimes I run my same tire in qualifying in the race. It depends, sometimes the surfaces are abrasive and you need a harder tire. Or sometimes I prefer to use a softer tire to get away a little sooner and then back off a little at the end and cool the tire back off.

"Qualifying for me is not that important. It's important to get a good starting position, but its more important to have the bike setup to the pace you want to run in the race. Myself, I've always managed to run a second-and-a-half quicker in the race than in qualifying. I just don't push myself that much in qualifying. It's good to have the bike fairly close in qualifying time to the race time because of the setup. Say you're three seconds faster in the race than in qualifying. There's a

The green light flashes and the riders accelerate toward the first corner, feet off the pegs, front wheels in the air for some, *straining to be in the lead at that first turn. Leading the first lap is good for the ego, but not necessarily good race strategy.*

good chance you'll crash in the race because you may have made the suspension too soft to handle the corners. You have to be running real close to your qualifying lap time in the race because it just drives off the corners differently."

As the weekend progresses—before the next timed session—the crew might freshen the engine. They might do a complete rebuild Friday night so that the rider has all new pistons for Saturday. That way they will have a broken-in engine before Saturday's timed session. "There's a progression," elaborates MacLean. "What throws a monkey wrench in there, of course, is if it rains, or if you have problems in the first timed session and then it rains in the second. Then you're in real trouble, because even though the bike is right, you can't go fast. The important factor that first day in that first session is to post a time so you know you'll be in the field. Then you begin to fine tune for the race."

The serious fans will have arrived by Saturday. But the teams will not have noticed. They will be continuing the fine-tuning, trying desperately to get to the head of the pack. They will be focusing on the tires, the suspension, the bumps on the track, the abrasion, the weather.

The *weather.*

Some rider seemed unphased by the hoopla surrounding each race. Here Doug Chandler manages to show his perpetual smile even when things are not going so well with the Cagiva.

The rider is saddled with confining leathers, pads, and safety gear which, like a fish out of water, makes him walk awkwardly. Yet on the machine, the gear is designed to make his job easier. Watching Doohan walk and then watching him move about the motorcycle during a race, one would think it was two different people.

In conditions that would make the average sedan racer cringe with trepidation, the Grand Prix rider considers it just another part of the contest. Rain plays a critical part of the weekend's strategy. Different riders and teams are affected by rain differently. Some riders aren't bothered by racing in the wet, others dread it. No matter what, if it rains, all the practice and all the adjustments go by the wayside. The teams must devise new set ups and new strategies. Then they pray . . . not for good weather, but for consistent weather. If they've been behind all weekend long, they hope for miracles.

The morning of the race is relatively uneventful. Warm-up is only twenty minutes for each class. If the bike isn't right yet, it never will be. By the same token, the mechanics can conclude their work, and that chapter of the season is closed as the bikes are pushed out to the pit lane for the final time of the weekend. The motorcycle is tuned in chunks of time, based on places—or rather racetracks. They are prepared for a time and place, and the race itself is basically won or lost on the first two days. Sunday (Saturday for the Dutch TT) is anti-climactic.

Eventually, the weekend is brought to fore. The riders mount their machines and set off for a warm-up lap before settling into their position on the grid. The rider is given a final look at the track with the sighting lap, and the pack finds the way back to the grid, where they will wait for the green light. Within seconds the green light will flash and the riders will accelerate toward the first corner, feet off the pegs, front wheels in the air.

The rider can plan a strategy to save his tires or suspension or engine—or even himself. There is definitely a plan. The race takes a toll on the bike. Mentally, an individual may not be able to win on a given day. If the rider sits atop an uncompetitive bike, but has conserved himself and the machine, he can capitalize on this should the front runners find trouble in the late stages. If he exhausts himself flogs the machine, he will pay later.

"I can tell how the performance has dropped off a little bit. At some places where the temperature and

At each race, gear is packed and shipped at the expense of the promoter or organizer of the next event. This Lufthansa cargo container holds equipment which left Italy four days earlier for the U.S. Grand Prix. Like a circus, this large, traveling show has loading and unloading down to a science.

The cargo container doubles as a dressing room, part-time office, and parts truck. Anything that can be loaded will be. More often than not, teams exceed the cargo allotment by three or four times.

the air are good, it might not ever change. But someplace where the air is a little thicker and it's not carbureted as good . . . It's just a feeling. It's just natural," explained Filice. "On a 500, it's really important—one or two jet sizes can mean a lot of horsepower. I have a hand in picking them, but most of that is the mechanic's guess. The air can change while you're out there. It's a weekend-long process. It starts on the first day by looking at the spark plugs, and then you have to compare notes from other days with that race and compare air density. It's gambling. Sometimes you lose and sometimes you win. That's why you try to pick the best mechanic you can because there are some guys who really know the bike and how far you can go down with the jetting and how to get the peak horsepower out of the bike."

Tracks vary from place to place, and a rider who excels at one venue may do poorly at the next. Laguna Seca played home to the U.S. Grand Prix 1988 to 1994 (with a one-year hiatus). To some, the place is difficult and dangerous, the perfect setup elusive. To others it is enjoyable and challenging. For Randy Mamola, Laguna is a special place. "One of the most difficult corners [at Laguna Seca]—in the world—is the Corkscrew, of course. The first time I went down it, I scared the shit out of myself. It was like, Jesus! I had always heard about it. I was about seventeen years old when I did it. But it's just a great feeling type thing. It's a fun part of the racetrack for me.

"When you go out, you're going down this straight in sixth gear. One of the toughest corners is turn one. On a 500 I think it's not frightening, but everyone's kind of cautions. It's like 160mph, but when the back end starts to spin and the front ends starts coming off

the ground coming over the hill, the object, of course, is to keep the power on and keep the tires spinning up over the top of the hill so it keeps the front end down so it drives it down the road. Turn two is basically a double left. It turns, and it has a little section where it's more straight then it turns back in on itself again. That is a corner where you can't spin it very much.

"If you stand here today you might be able to see them spin coming out of turn eleven. Another corner [to watch for spinning] would be the outside of turn four. You gotta stand somewhere where you can see them coming out of the corner from behind. If you can get up above to look down on them, you'll see them there too. Getting up the hill takes a lot of horsepower, and the faster you can get around that hill at the bottom, the faster you're going to get up the hill. It's one of the most difficult corners because it's a blind corner. The other corner I like is turn ten coming into the pits. It's not a track where you're tying to spin the bike a lot—this corner, turn eleven, turn four, turn three a little bit, turn five a little bit, turn six none at all."

Grand Prix racing is an infinite series of adjustments culminating in, hopefully, the optimum setup for a particular track. The engineers can alter the wheelbase, *the steering head angle, the carburetor settings, power valves, and a plethora of other things.*

Danny Walker, who occasionally leases bikes for one-off Grand Prix events, has been in the enviable position of being able to take an amateur's ride on a pro circuit. The riders who choose to do only one Grand Prix a year, as Walker did, need to qualify within 110 percent of the fastest rider (for example, if pole is 60sec, an amateur must turn 66sec or better). Essentially, the newcomer has to jump in and not exceed Kevin Schwantz' or whoever's pole time by more than

10 percent. And even if he manages to make it, he's still got to keep the pace throughout the race, which Walker found, was quite a challenge.

"It's quite a bit different style in riding," recalled Walker. "A lot of the guys have been saying that throughout the years. Things happen a lot quicker on a 500. Your depth perception, your speed perception all has to go through a major change. They're easy to ride around the track. They ride like a street bike, but

Some riders prefer to ride to work. Here Kevin Schwantz moves the Lucky Strike Suzuki RGV Gamma out to the track for the first practice.

Pit boards provide the rider an array of information while he is on course. Usually it will tell him his position and time ahead or behind the next competitor. This one, held by Team France for Freddie Spencer, was photographed Friday at the 1993 Grand Prix. It is only being shown to get Spencer conditioned to where to look; there is no message on the board.

when you want to go fast, you've got to be really careful. It'll spit you off real fast."

The elite are defined as such because of how they control the motorcycle on the racetrack. "When Kocinski came by me in practice, the guy was on the edge. The bike was sideways; the front wheel was coming off the ground mid-corner. It's shaking, wobbling, and his feet are flailing off the pegs. I mean, he was going. You really want to see if you can do it or not. That's the elite—that's what you shoot for. The biggest thing that surprised me was how slow everybody went at the beginning of the race. They really weren't going that fast the first couple of laps; they really seemed to be waiting until everybody got settled before they really started racing."

Not everyone is banking on a good finish, and racing is not only about winning. Back at the motorhome, the race takes on a different feel, a different meaning. "The function of an owner is to assemble a team which has the ability and professionalism to not only perform well but to have the professionalism and ability to attract corporate sponsorship," says MacLean. "Corporate sponsorship is what we're really taking about—obviously, money. Your team should reflect the image of the corporate sponsor. It should be well put together, it should be well-managed, it ought to be fast, and whatever all those things the cor-

porate sponsor feels is part of its image. It is the responsibility of the owner to assemble the package and that includes everything from how you travel, where you stay, how you appear in public, what you say to the press, what the rider's leathers look like, what the truck looks like, does the mechanic have insurance, are they being fed? Your interest is in the whole. The whole is effected by all of those things and the race is only apart of the whole. Not everybody can win. There's only one winner of a given race. Why are all those companies in there? If they felt that they just wanted to win, then there wouldn't be any sponsors—everybody would go away. There can only be one winner. So there must be lots of other things that are important to the sponsor. The arena that they compete in is important. Do they want Grand Prix, AMA, do they want dirt motocross, do they want Formula One? What I am possessed with is the whole thing—how we relate to people, how we relate to people on the track, how we relate to people off the track, all of those things, and the race is just part of it.

"My role—the things that I do—should have already been done by the time we get to the racetrack. The corporate things. Finding out who's coming to

Grand Prix racing is popular worldwide, with support from twelve countries which host Grands Prix. However, the exposure and following is global, from Macau to Alaska. Here, an autograph session with Honda riders attracts a huge crowd, with the line extending a nearly a half mile for the chance to meet Mick Doohan or Shinichi Itoh.

Free practice starts at 9:00 AM on Friday for 250s and 10:15 for 500s. Practice will last an hour. The rider changes into his leathers and wanders out to pit lane. The crew will have wheeled the machine to the front of the garage. As the clock approaches the appropriate time, the bike, now ready, is pushed and bump started. It coughs to life and belches the first cloud of sickly-sweet blue smoke. Moments later the next bike is jerked to life, then another and finally the entire pit lane is a cacophony of engine song.

the race, making sure they have rooms, making sure that if we're going to meet a potential sponsor that we meet them and figure out where they're staying. Obviously I'm interested in the racing, and I want to know what's going on there, too."

After an all-too-short hour-long race, it's over. The next round will begin immediately—in fact, on the cool-down lap. The engineers will begin discussing what went wrong and what went right. Did the rider get the best machine the team could provide? Did the team get the rider's best performance? Suggestions will be made and perhaps new innovations will be elicited. It will be better the next time.

Hopefully, there were no injuries or major problems. One rider will have captured twenty points toward the championship. Fifteen riders will have scored some points. Most likely, things turned out as they were supposed to have. Racing is a brutal sport, where anyone can win given some luck. But preparation increases the odds. However it turned out—no matter how one sided it may have been—it was significant as one chapter of the season-long competition.

"The race is almost an afterthought," said MacLean. "We almost know what's going to happen. When you've got the three top guys that are very close, and they're young guys that are really going for it, then you have all the unknowns of aggression and chance and all of those kinds of things. The whole build-up is probably more draining in a way than the race itself.

"You know if nobody falls off and if nothing breaks, I can probably tell you on that given day where we're going to finish. You can tell by looking at the time sheet, you can also factor in reliability, and you know that the guys in front of you will have something fall off the bike and so on. In a completely separate context, the race, like the grade on the exam you took in school, legitimizes all the work you've done. It's, Look at me, after all the work I've done I was ninth in the world, or whatever. Well that's pretty damn good and everybody feels pretty good about it."

Subtle changes mean the difference between winning and losing: "On a 500 jetting is really important," said Jimmy Filice. "One or two jet sizes can mean a lot of horsepower. I have a hand in picking them, but most of that is the mechanics guess. The air can change while you're out there. It's a weekend long process. It starts on the first day by looking at the spark plugs and then you have to compare notes from other days with that race and compare air density. It's gambling. Sometimes you lose and sometimes you win. That's why you try to pick the best mechanic you can because there are some guys who really know the bike and how far you can go down with the jetting and how to get the peak horsepower out of the bike."

Riders can choose to pace themselves and save the tires, machinery, or just themselves. There is a large degree of planning, and it is just not the fastest man on the day that takes the trophy. Riding hard puts a toll on the bike; working the suspension and tires hard may leave a rider with the talent and strength to win, but without a competitive bike.

INDEX